STRENGTH OF DILATING SOIL AND LOAD-HOLDING CAPACITY OF DEEP FOUNDATIONS
INTRODUCTION TO THEORY AND PRACTICAL APPLICATION

Strength of Dilating Soil and Load-holding Capacity of Deep Foundations

Introduction to theory and practical application

DMITRY YU. SOBOLEVSKY
Minsk, Republic of Belarus

A.A.BALKEMA/ROTTERDAM/BROOKFIELD/1995

Authorization to photocopy items for internal or personal use, or the internal or personal use of specific clients, is granted by A.A.Balkema, Rotterdam, provided that the base fee of US$1.50 per copy, plus US$0.10 per page is paid directly to Copyright Clearance Center, 222 Rosewood Drive, Danvers, MA 01923, USA. For those organizations that have been granted a photocopy license by CCC, a separate system of payment has been arranged. The fee code for users of the Transactional Reporting Service: 90 5410 164 4/95 US$1.50 + US$0.10.

Published by
A.A.Balkema, P.O.Box 1675, 3000 BR Rotterdam, Netherlands (Fax: +31.10.4135947)
A.A.Balkema Publishers, Old Post Road, Brookfield, VT 05036, USA (Fax: 802.276.3837)

ISBN 90 5410 164 4

© 1995 A.A.Balkema, Rotterdam
Printed in the Netherlands

*Dedicated to
Larisa*

Contents

Introduction   1

Part 1: Strength of dilating non-cohesive soil

1 Dilatancy as a fundamental property of granular media   5
   1.1 On the term 'dilatancy'   5
   1.2 Interaction of grains in non-cohesive soils   5
   1.3 Changes in soil density in the process of failure   9
   1.4 Plastic and elastic deformations of non-cohesive soils   12
   1.5 Dilatancy and the stress condition of soil   13
   1.6 Examples of free and constrained dilatancy   16
   1.7 Manifestation of dilatancy in soil strength tests   18
   1.8 Conclusion   21

2 Models of elasto-plastic deformations of dilating non-cohesive soil   23
   2.1 General propositions   23
   2.2 Models of contact shear   24
   2.3 'Soil-to-soil' shear model   27
   2.4 Model of internal bulge   29
   2.5 Justification of the assumption regarding elastic reaction of the massif to dilatancy   31
   2.6 Conclusion   32

3 Instruments and methods for soil testing in the conditions of constrained dilatancy   34
   3.1 General propositions   34
   3.2 Dilatometric Apparatus of Contact Shear (DACS)   34
   3.3 Dilatometric Apparatus of Direct Shear (DADS)   39
   3.4 Special testing method with the use of the serial shear apparatus BCB-25   41

Contents

3.5 Dilatometric Apparatus of Contact Shear of reinforcing elements
     (DACS-A) 42
3.6 Dilatometric Triaxial Apparatus (DTA) 44
3.7 Conclusion 48

4 Contact shear in conditions of constrained dilatancy 49
   4.1 General propositions 49
   4.2 The influence of dilatancy constraint on resistance of soil to
        contact shear 51
   4.3 Dilatant component of strength as a function of massif elasticity ... 56
   4.4 Angle of contact friction as a function of massif elasticity 57
   4.5 The influence of initial soil density and grain size 59
   4.6 Influence of moisture content 63
   4.7 Values of dilatant strains 65
   4.8 Influence of grain strength 69
   4.9 Critical density and critical normal pressure 69
   4.10 Peak and residual strength of sands 71
   4.11 On reasons for curvature of function $\tau_u = f(\sigma_{n_o})$ 72
   4.12 Cyclic shear in conditions of constrained dilatancy 73

5 Direct shear in conditions of constrained dilatancy 75
   5.1 General propositions 75
   5.2 Influence of constraint on dilatancy on resistance to shear .. 77
   5.3 Dilatant component of strength and angle of internal friction as
        functions of massif's elasticity 78
   5.4 Comparison and peculiarities of tests in dilatometric instruments
        of various designs 80
   5.5 Conclusion 84

6 Internal bulge as a manifestation of conditions of constrained dilatancy ... 86
   6.1 General propositions 86
   6.2 Influence of dilatancy on stress-deformative condition of soil
        during triaxial compression 87
   6.3 Adjustments for dilatancy in Coulomb-Mohr's strength conditions ... 87
   6.4 Angle of internal friction as a function of massif elasticity ... 91
   6.5 Peculiarities of soil deformation with constrained dilatancy . 92
   6.6 Conclusion 94

7 Conditions of strength of dilating non-cohesive soil 96
   7.1 General propositions 96
   7.2 Ultimate resistance to shear 96
   7.3 Ultimate state during triaxial compression 101
   7.4 Conclusion 106

Part 2. Deep foundations in dilating soil

8 Constrained dilatancy as a factor of load-holding capacity of deep
   foundations   111
   8.1 General propositions   111
   8.2 Bore piles of Type 1   113
   8.3 Bore piles of Type 2 and footings of Type 1   115
   8.4 Bore piles of Type 3   119
   8.5 Bore piles of Types 4, 5 and footings of Types 2, 3   124
   8.6 Injection piles and anchors   131
   8.7 Piles constructed with soil displacement   135
   8.8 Factors of stress-deformative condition at the contour of a bore pile 138
   8.9 Conclusion: Reasons for the failures of theoretical calculation
        methods   145

9 Load-holding capacity of a single pile   150
   9.1 Calculating scheme   150
   9.2 Propositions regarding calculation of bore piles   150
   9.3 Propositions for calculating bore piles with injected base and shaft   158
   9.4 Propositions for calculating injection piles   161
   9.5 Propositions for calculating piles manufactured with displacement of
        soil   162
   9.6 Conclusion   165

10 Load-holding capacity of a deep footing   167
    10.1 Calculating scheme   167
    10.2 Load-holding capacity along the skin surface   167
    10.3 Load-holding capacity at the lower end   170
    10.4 Equation for calculating the total load-holding capacity   171
    10.5 Conclusion   172

11 Load-holding capacity of an injection anchor   173
    11.1 Calculating scheme   173
    11.2 Contact resistance to shear along the root surface   176
    11.3 Critical length of compressed root   180
    11.4 Equation for calculating the load-holding capacity   182
    11.5 Group effect of anchors   183
    11.6 Conclusion   185

12 Sketches of several dilatancy manifestations   186
    12.1 Reinforced earth as a composite material   186
    12.2 Distortion and liquefaction of sands   195
    12.3 Dilatant nature of contact filtration and negative friction along the
          pile shafts   200

12.4 Realization of the factor of dilatant strength during tunnelling   202
12.5 Dilatancy and seismic activity   203
12.6 Dilatancy in cracks of stone constructions   204
12.7 Reinforced concrete as a dilating composite material   206

General conclusions   208

Appendix 1   211

Appendix 2   225

References   237

Subject index   243

Introduction

Modern geotechnique raises a number of questions about soil mechanics, and many of these questions have still not been satisfactorily answered. Primarily, it bears on the load-holding capacity of deep foundations. Quite often a comparison of experimental data of resistance to contact shear and internal bulge with those calculated theoretically yields, contradictions which appear inexplicable within the framework of traditional strength notions. These contradictions have come to a head in connection with the expansive introduction into practice of new technologies which ensure, on the one hand, minimal violation of natural soil properties, and on the other hand, active and purposeful ameliorative influence on the soil. These methods are the trenchwall method, injection compaction of bore piles' bases, injection anchoring, soil reinforcement etc.

The theory seems to be lagging behind, not only in working out methods for calculating bearing capacity and stability, but also in giving a theoretical explanation of such methods and technologies, namely the physical foundation for their influence on soil properties. What we have here can be described in terms of a crisis, manifesting itself in a wide spread of empirical and semi-empirical approaches. This has given rise to a justified scepticism among practising engineers as to the assessment of the validity of traditional propositions of strength and deformation of soils.

In this connection it would be useful to remember that the strength theories currently employed by soil mechanics were originally worked out for solid bodies. Realisation of this fact has caused numerous attempts to refine and amend these theories. Special testing techniques have been developed. A number of suppositions were made as to the connection between strength parameters and conditions of deformation, but the material accumulated so far has not been theoretically generalized as yet, even if it constitutes an important contribution to science. An introduction of numerous amendments, while complicating the existing theories, allows to grasp the consequences rather than establish the reasons for physical differences in the descriptions of behaviour of soil and solids.

At the same time a fundamental property exists which is not characteristic for solids and liquids and which is inherent for a granular medium. This is the property of changing volume in the process of form deformation, which is how it was originally defined by Reynolds in 1885. He termed this property 'dilatancy'.

Over a century has passed since the property of dilatancy was discovered. During this time the awareness of soil interaction mechanisms has increased, and the realization of internal friction has come to be considered as the source of strength in non-cohesive mediums. However, it would be an exaggeration to state that the fundamental property of dilatancy has been fully represented in soil mechanics, especially in its use for the practical needs of geotechnics. No tangible advance is observed in the development of strength and soil deformation theories, either. Dilatancy is de facto ranked as a minor phenomenon which is of interest to soil scientists rather than to engineers.

This work is a study of dilatancy as the possible key to removing some of the accumulated backlog of contradictions between soil mechanics and practical geotechnics.

1. Strength of dilating non-cohesive soil

CHAPTER 1

Dilatancy as a fundamental property of granular media

1.1 ON THE TERM 'DILATANCY'

The term 'dilatancy' was introduced by Reynolds (1885). He drew the attention of researchers to the property of granular media which is not possessed by known 'fluids or solids', namely, the property to change one's volume in the process of change of shape or distortional strain.

Reynolds found out during his experiments that in the process of plastic deformation, dense granular mediums (sands, densely packed glass balls) increase their volume, dilate, whereas loose mediums, on the contrary, contract, decrease their volume. If the granular medium was saturated with liquid, the dilatancy was accompanied by a decrease in the medium's humidity, while in the case of contraction the liquid was squeezed out of the pores.

In 1886 Reynolds described experiments with water-saturated samples and proved that in cases where drainage is limited, dilatancy causes a steep drop of pressure in pores. Reynolds research led him to the conclusion that dilatancy is 'a singular fundamental property of granular media' which need a special consideration and description. At present the term 'dilatancy' is often used to designate any type of volume change in the conditions of distortion. If so, contraction may be defined as negative dilatancy.

1.2 INTERACTION OF GRAINS IN NON-COHESIVE SOILS

Ideally, non-cohesive soils can be defined as soils, wherein the interior forces or intergranular connections influence the mechanical behaviour only insignificantly (Bishop, 1975). Soils falling under this definition present a grainy structure where the interaction of separate grains is effected by means of contacts along their facets. The number of contacts in a unit of volume is determined by the grain size and the density of their packing. The pressure at contact points from the stresses effective in the soil is determined similarly.

The equilibrium of the grainy medium is ensured by the forces of mineral

6 *Dilatancy as a fundamental property of granular media*

(intergranular) friction and mutual grain gear. The mechanism of mineral friction according to Lambe & Whitman (1969, 1977) is represented in Fig. 1.1a as friction of grains of the same size against a smooth plate made of the same material. The value of ultimate resistance to shear τ_u measured during such an experiment is defined as

$$\tau_u = \sigma_n \operatorname{tg} \varphi_\mu \tag{1.1}$$

where σ_n is normal pressure at the shear surface; φ_μ is the angle of the so-called mineral friction.

Actually, in real soils the mechanism of mutual grain shear is much more complicated. As was demonstrated by Rowe (1975), the grains make closed groups which at each given moment slide one against the other(s) with consequent formation of new groups. The sliding contacts are instantaneous and involve only a limited number of grains. Rowe's illustration (1975) for such contacts is represented in Fig. 1.2. This treatment of the mechanism of grain sliding warrants the denomination of φ_μ as the angle of intergranular friction, which appears more justified.

Closed grain groups experience gearing during the mutual shear, and it is this gear that prevents the sliding contacts from becoming ordered. The example provided by Lambe & Whitman (1977) and represented here in Fig. 1.1 b, c which treats the mechanism of grain gear with dense and loose grain packing, is also relevant here.

Summing up the results of their research on soil shear, Lambe & Whitman (1969, 1977) came to the following conclusions:

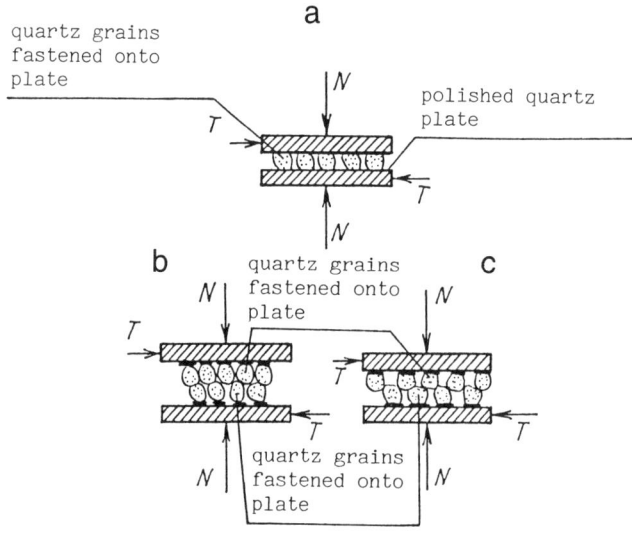

Figure 1.1. Mechanisms of: a) mineral friction; b) mutual gear with slightly jagged surface; c) mutual gear with strongly jagged surface (Lambe & Whitman, 1977).

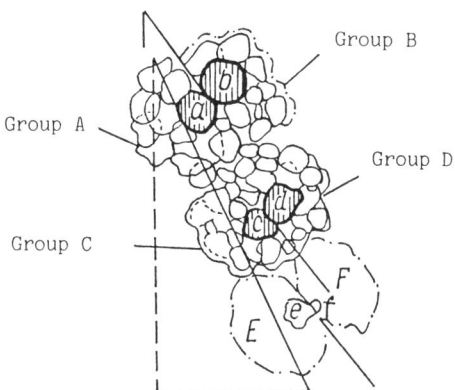

Figure 1.2. Instantaneous distribution of sliding contacts with shear (Rowe, 1975).

1. The higher the density of sand, the larger the mutual grain gear and the value of the angle of internal friction;
2. The higher the density of sand, the larger the increment of sand volume with shear;
3. The resistance of sand to deformations decreases parallel to the increase of the sand's volume;
4. The decrease of resistance to deformations is most typical for high porosity samples.

According to Rowe (1975), when the grain packing is loose, the slidings develop throughout the deformation area in several directions simultaneously, i.e. the shear takes place in larger volumes of soil.

As applied to the function (1.1), the grain gear causes an increment of the angle of intergranular friction φ_μ by the so-called angle of gear φ_g. When added up, these values form a strength parameter called the angle of internal friction φ, i.e. under this treatment

$$\varphi = \varphi_\mu + \varphi_g \qquad (1.2)$$

The angle of intergranular friction conforms better with the grain forming mineral and the chemical properties of contact. The latter are determined by the prehistory of grain-formation and the presence of liquids in the sliding areas (Rowe, 1975). Hardy & Bircumshaw (1925) also proved that specific pressure accounted for by the grain contact has a certain influence too. With the increase in pressure, though φ_μ, decreases down to a certain constant value.

The measurements of angles of intergranular friction in grainy mediums, taken by different authors and summed up by Rowe (1975), are represented in Table A1.3. The values of φ_μ for quartz grains are calculated in the range 23-30°, with mean values for sand equalling 26-28°. According to Goldstein (1979), the angles of mineral friction of quartz grains sliding along the smooth surface of a quartz plate constitute 24-28° (Table 1.1).

8 Dilatancy as a fundamental property of granular media

Table 1.1. Coefficients of mineral friction $tg\varphi_\mu$ and angles of mineral friction φ_μ, grades for smooth grains (according to Goldstein, 1979).

Mineral of the skeleton	Dried at 110°C		Dried at 100°C*		Water-saturated	
	$tg\varphi_\mu$	φ_μ	$tg\varphi_\mu$	φ_μ	$tg\varphi_\mu$	$\varphi_\mu\mu$
Quartz**	0.13	7.4	0.13	7.4	0.45	24.2
Field spar	0.12	6.8	0.12	6.8	0.77	37.5
Calcite	0.14	8.0	0.14	8.0	0.68	34.2
Muscovite	0.43	23.2	0.30	16.7	0.23	13.0
Flogopit	0.31	17.2	0.25	14.0	0.15	8.5
Biotit	0.31	17.2	0.26	14.6	0.13	7.4
Chlorite	0.53	27.9	0.35	19.3	0.22	12.4

* With consequent air exposure.
** For very coarse quartz grains $tg\varphi_\mu = 0.5$ ($\varphi_\mu = 26°$).

It would not be incorrect to state that for a given soil φ_μ is a relatively constant parameter, whereas the angle of gear φ_g is a variable depending on roughness, density of packing, and granulometric composition.

Resistance of grainy medium to shear obeys Coulomb's law of dry friction:

$$\tau_u = \sigma_n \, tg \, (\varphi_\mu + \varphi_g) \quad (1.3)$$

The function (1.3) presupposes that the maximal shearing stress on the one hand, and the normal pressure effective in the shear plane on the other hand, are in direct proportion (Fig. 1.3). Similarly, according to Rankine and taking into account Mohr's strength condition, the strength of soil is regulated by the relationship of the biggest σ_1 and the smallest σ_3 of the principal stresses:

$$\frac{\sigma_1 - \sigma_3}{\sigma_1 + \sigma_3} = \sin \varphi \quad (1.4)$$

Providing the conditions (1.3) and (1.4) are correct, we must integrate into the parameter φ such factors as roughness, size and mineral composition of grains, density of their packing, humidity. The Hansen-Landborne equation is an attempt to represent the influence of these factors on the value of φ. According to this equation,

$$\varphi \approx 30° + \Delta\varphi_1 + \Delta\varphi_2 + \Delta\varphi_3 + \Delta\varphi_4 \quad (1.5)$$

where: $\Delta\varphi_{1-4}$ are adjustments (see Table 1.2).

After initial acquaintance with the function (1.3), the parameter φ appears invariable relative to the failure conditions, or as a certain constant value for the soil of given density and composition, and for the liquid saturating its pores. But this analogy with shear conditions when solid bodies are in contact has to be considerably amended due to the dilatancy phenomenon, which, it will be remembered, was defined by Reynolds as a singular fundamental property of granular medium. We will now turn to the analysis of these amendments.

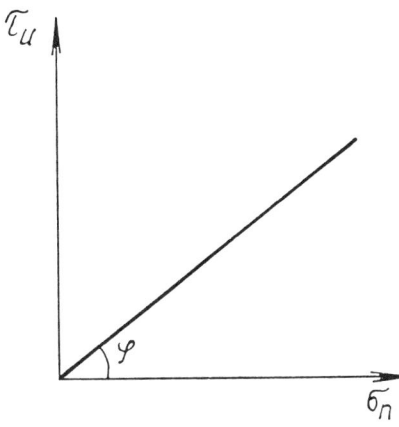

Figure 1.3. Coulomb's function for non-cohesive grainy medium shear.

Table 1.2. Values of adjustments $\Delta\varphi_1$-$\Delta\varphi_4$, degrees.

Factor	Soil characteristics	Value and polarity of adjustment, degree
Roughness of grains $\Delta\varphi_1$	Rough	+1
	Medium roughness	0
	Rolled	–3
	Well rolled	–5
Grain size $\Delta\varphi_2$	Sand	0
	Fine gravel	+1
	Medium and coarse gravel	+2
Non-homogenety of granular composition $\Delta\varphi_3$, with the non-homogeneity index $C_u = d_{60}/d_{10}$	Homogeneus soil $C_u \leq 3$	–3
	Non-homogenous soil $C_u \geq 3$	+3
Density of packing $\Delta\varphi_4$	Highly loose packing	–6
	Medium density	0
	Highly dense packing	+6

1.3 CHANGES IN SOIL DENSITY IN THE PROCESS OF FAILURE

The failure of non-cohesive soil always presupposes a change in the original grain packing, which takes place in the process of overcoming the forces of intergranular friction and gear, i.e. is expressed in shears. Observation reveals the fact that the majority of soils (with the exception of very loose ones) reveal an increase of volume in the process of failure under not high normal pressures (0-300 kPa). Loose and very dense soils under high normal pressures in the process of failure contract.

Let's use Lambe & Whitman's example (1977) to illustrate the process of dilatancy. In Fig. 1.4a, a system of evenly packed balls represents a maximum

10 *Dilatancy as a fundamental property of granular media*

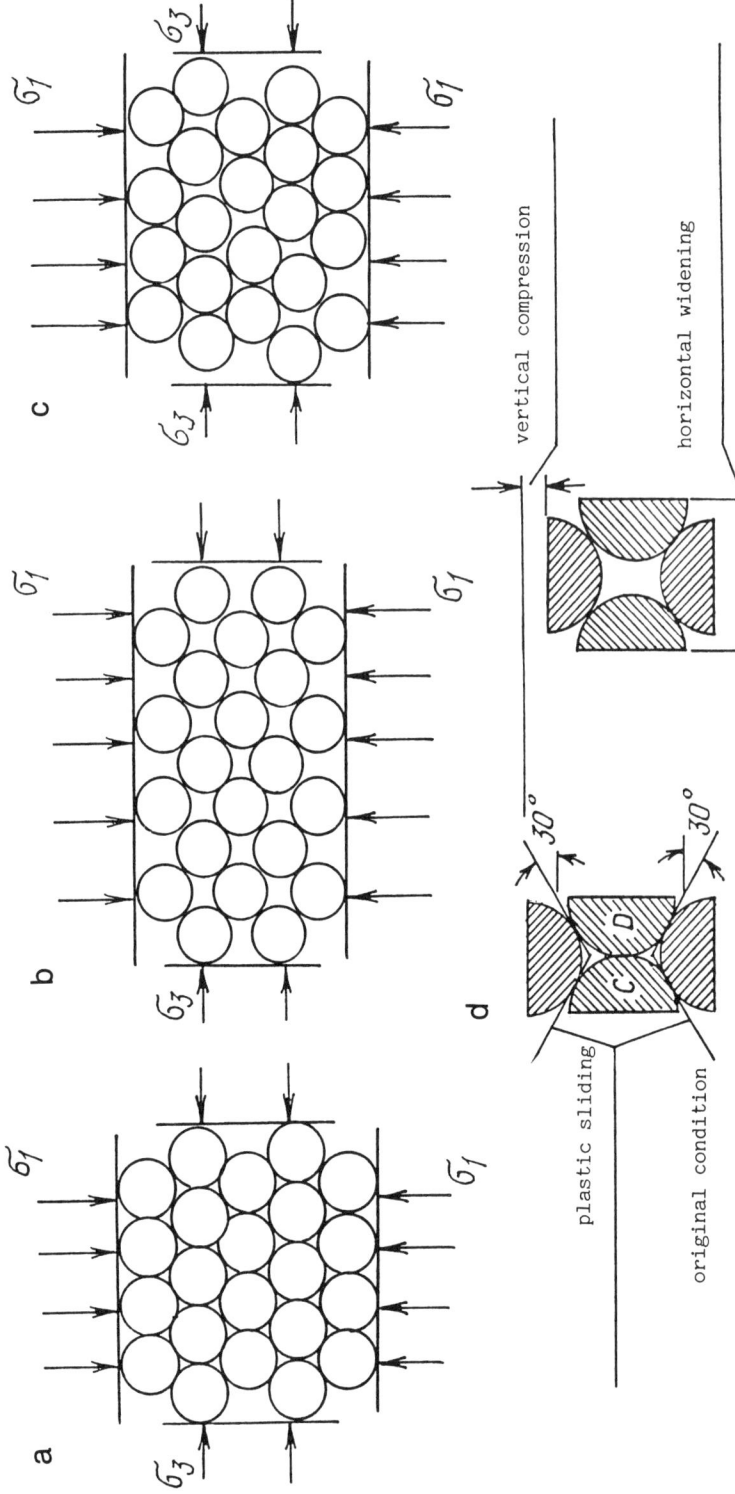

Figure 1.4. Deformation of regular grain packing: a) original packing; b) maximally loose composition with regular packing; c) loose condition with uneven packing; d) change in volume of an exemplar grain group (Lambe & Whitman, 1977).

dense packing of soil in a certain initial stress state. The increase of the principal stress σ_1 causes a decrease in density as a result of a mutual shear of the balls to the state, which is as loose as possible for a regular packing (Fig. 1.4b). On the contrary, a lateral compression of the system causes a loosening with a breaking of the regular packing (Fig. 1.4c).

Evidently, there must exist a certain intermediate value of density, with which the volume of soil remains unchanged both prior and consequent to the failure. This value was termed 'critical density' (porosity) (Casagrande, 1936).

The concept of critical density was the first step to the quantitative consideration of dilatancy in the mechanics, and it gave an impetus to the creation of the concept of critical condition, which was defined by Hvorslev (1937) as follows: 'Finally a condition is achieved, under which further deformation does not lead to changes in resistance to shear and porosity coefficient. The final values of these quantities are equal in both cases (i.e. in loose and dense conditions) and, consequently, do not depend on the initial value of soil density'.

Casagrande thought that the density of sands always reaches the critical value in the process of shear. This has been a moot point for a long time. The discussions thereof gave birth to the term 'normally dense condition', which characterizes the absence of volume deformations with an continuous shear.

Nevertheless volume deformations of soil in the process of failure are, as a rule, not taken into account during strength tests. At the same time the question arises: to what density of soil must the measured strength parameters be referred – the original density or the normally dense condition? If the latter is the case, it appears outwardly that the strength parameter must principally depend on granulometric and mineral composition of grains and their roughness, whereas the original density of packing, structure and texture seem to be allegedly minor factors. But one must not forget that the kinematic conditions of dilatancy or of the soil reaching the critical density condition are really determined by these 'minor' factors.

The method of interpreting soil strength tests, taking into consideration the dilatancy of soils, or coming to the critical density condition, was worked out by Ryzhov (1976). He also suggested an original method of taking into account the potential soil deformations occurring in the process of failure. As applied to the traditional types of soil tests, Ryzhov's method of dilatancy awareness appears to be the most accomplished and methodical.

Critical density is a quality which characterizes a transition of soil from loose to dense packing. As it is demonstrated by the research conducted by Rowe (1975); Dresher & de Jong (1972); Ryzhov (1976) and other authors, the density of grain packing determines the character of their interaction in the process of failure. In the case of loose soils the repacking of grains encompasses considerable volumes, as a rule, i.e. general failure takes place. In dense soils the failure is localized. The grain shear occurs within a narrow band of the surface of sliding (rupture). Outside this band no grain repacking takes place. The soil 'splits' into separate blocks which shear relative to each other (Ryzhov, 1976). As Nikolaevsky points out (1975), 'failure and the formation of lines (planes) of rupture are much more probable in dense soils'. And finding the lines of sliding becomes a major problem when calculating ultimate equilibrium of earth massifs.

1.4 PLASTIC AND ELASTIC DEFORMATIONS OF NON-COHESIVE SOILS

The change in the density of non-cohesive soil is a result of plastic deformation ensuring from the repacking of grains, which can be expressed in their mutual sliding, turning or splitting off. The sliding is typical for dense soils; the turning and sliding accompanied by a reorientation of grains are more characteristic for loose soils. We will disregard the factor of deformation of water films at contact points in our further consideration, keeping in mind the fact that moistening during the sliding influences the values of the angle of intergranular friction φ_μ (Rowe, 1975).

The character of grain movement depends on the degrees of freedom allowed by the packing pattern, and on the forces of intergranular friction. As Rowe demonstrated (1975), in the case of dense packing no grain-turning is observed, as a rule. In his shear experiments, where glass balls imitated soil, the turning of the balls was registered only in the case of regular packing, when one ball was free and was not loaded.

The experiments conducted by Rowe (1975); Dresher & Jong (1972) explain why the nature of failure is different in the cases of dense and loose grain packing. In the former case mutual displacement of grains is extremely constrained by forces of intergranular friction at contact points, and the slidings are localized only in the planes of highest shearing stresses. The repacking of grains with their possible turning encompasses only a thin layer along the plane of sliding, where the loosening to the critical density condition occurs. In the adjacent soil layers the shear causes the formation of rigid grain chains, between which sliding takes place very seldom, if at all. As Ryzhov (1976) points out, 'porosity almost does not change outside the shear zone'. Further, if no irreversible mutual displacement of grains takes place in the soil, it would be correct to state that the soil is undergoing an elastic deformation. We can conclude, that in dense soils plastic deformations and, consequently, dilatancy are localized along the sliding surfaces.

The phenomenon of grain split off becomes essential when very high stresses develop in soil. Lambe & Whitman (1977) measured them at 700 and more kPa for quartz sands, which is a rather infrequent occasion in geotechnics.

The thickness of the sliding band depends on the density of soil packing, the soil's granulometric composition and roughness. The author did not have a chance to come across special research to this effect, but in a number of cases the

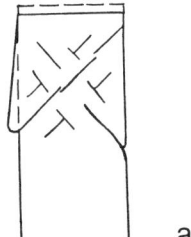 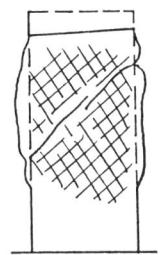 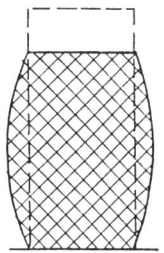

Figure 1.5. The character of failure of soil samples with broken structure depending on the density of packing: a) overcompacted condition of soil; b) normal compaction; c) undercompacted condition of soil.

measurements of the zones where shear deformations concentrate were taken parallel to other research. For instance, Bishop (1975) gives the measurement of thickness for medium-sized soil volumes equal to 2-3 mm, mentioning that outside this layer soil volumes are deformed as blocks. Wernick (1977) correlated this value with mean grain diameter d_{50} and found out that it constitutes approximately $15d_{50}$ for dense sands with contact shear. Goldstein (1979) states that 'apparently the thickness of the shear zone is bigger for loose sand, and smaller for dense sand'.

No rigid constraint on grain movements freedom is observed in loose soils. Sliding occurs simultaneously in several directions (Rowe, 1975). Better conditions for grain gear ensure the possibility for the grains to turn, which, strictly speaking, decreases the possibility for starting the mechanism of intergranular friction. In other words, not all the grains, drawn into the zone of plastic deformation, experience sliding. This causes the drop of the φ_μ component in the value of mobilized angle of internal friction. As a result, one can observe failure as a plastic flowing, as shown in Fig. 1.5.

It is worthwhile noting that in natural conditions plastic deformation in considerable volumes of soil not always presupposes its loose packing. The character of soil deformation is determined not only by the original density, but also by the stress condition and the conditions of failure. For example, during the so-called internal bulge under the lower end of a pile, even dense soil is subjected to general failure in considerably large volumes.

1.5 DILATANCY AND THE STRESS CONDITION OF SOIL

As Kenny (1967) pointed out, strength in the process of shear must be regarded as the behaviour of soil, not its property, and this behaviour depends both on the values of characteristics of strength and on the stress condition. Changes of this latter are capable of considerably influencing the strength.

In its turn, taking into account the changes in stress condition means introduc-

14 *Dilatancy as a fundamental property of granular media*

ing into the calculation the deformational characteristics of the material, and, consequently, there must be a correlation between the strength and deformational parameters of soil. Goldstein (1979) pointed out that such correlation was completely denied but recently 'because it appeared that various molecular processes lay at the basis of such properties as elasticity on the one hand, and resistance to failure on the other hand'. But it is really intergranular interaction rather than intermolecular processes that is of importance when we deal with disperse materials, such as soil.

Research demonstrates that critical density, which can be a criterion for failure of soil, is inseparable from its stress condition. For example, when even dense sand experiences shear, the measured dilatancy decreases with the increase of normal pressure, and at certain values of the latter dilatancy can be substituted by contraction (Fig. 1.6). But both with low and extra-high pressure in the process of shear, the condition of critical density will be achieved.

It appears necessary to introduce the quantity 'initial critical density with a given level of pressures'. This quantity corresponds to the case when the volume of soil before and after the failure remains the same. This quantity for soils presupposes, that dilatancy is localized on the sliding surface.

The suppression of dilatancy with the increase of normal pressure, registered in shear experiments, presupposes a corresponding thinning of the grain layer involved in repacking. This can be explained physically, because the higher the pressure at the contact points is, the harder it is to cause sliding (according to (1.1)).

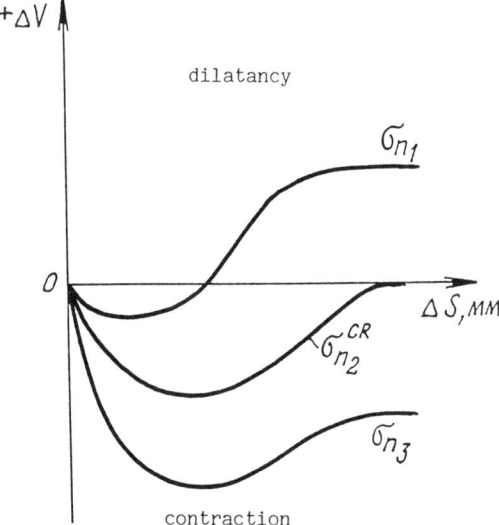

Figure 1.6. Dilatancy of dense non-cohesive soil in the process of shear: when the normal pressure surpasses the critical value, it causes contraction.

The dilatancy emerging in the shear zone wedges out the adjacent layers of soil. If the wedging effect is not limited in any way, the dilatancy does not influence the stress condition in the process of shear. But when dilatancy is limited it becomes a significant factor of the stress condition.

To clarify this, we shall take the case of thrust of soil from under a shallow foundation, and the so-called internal bulge under the lower end of a pile (Fig. 1.7a, b). In the first case the dilatancy, which develops in the thrust area, is not limited and is not capable of influencing the resistance to shear in any significant way. In the second case the area of plastic deformations is within the soil massif which constrains dilatancy and, consequently, causes the developments of additional stresses. These stresses naturally cause an increase in the process of failure of stresses at grain-contact points, thus increasing their resistance to mutual repacking.

Reynolds pointed out the possibility of the influence of constrained dilatancy upon the strength of the grainy medium back in 1885. As an example he gave the process of loading two sacks with sand. One sack was made of soft caoutchouc, the other of hard rubber. The first sack very quickly ceased to resist and sprawled down, whereas the second sack could withstand considerable pressures. The final density and porous pressure turned out to be different in the two sacks. This

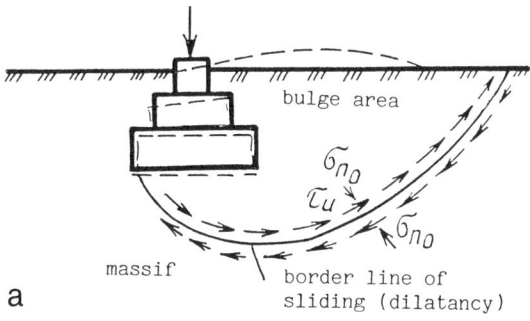

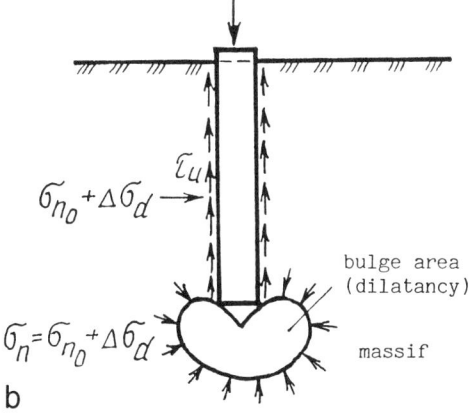

Figure 1.7. External and internal bulge: a) from under a shallow foundation – normal pressure along the outer sliding line during shear is constant at all points; b) from under the lower end of the pile – dilatancy limits the thrust due to the reaction of the massif.

experiment clearly demonstrates that deformability of soil in the failure area depends on the dilating conditions granted by the environment surrounding this area. This is the case when the correlation between strength and deformability of soil, which was tentatively mention by Goldstein (1979), begins to work. In the case of internal bulge the part of the elastic bag is played by the soil massif. It is evident that resistance to thrust will be increased with the increase of resistance of the massif to the volume increment in the thrust area. In other words, the elastic deformation of the massif will determine the condition for the development of plastic deformation. Similarly in the process of contact shear along the shaft of the pile dilatancy experiences the reaction of the surrounding massif, and this reaction determines the values of the mobilized normal pressures and the condition for the shear of grains in the sliding band.

Tschebotarioff (1968) noted that if loosening of dense sands can be prevented, 'the consequence of this would be the increase of the limiting resistance to shear to unexpectedly high values'. Ivanov (1985) explained this fact in the following way: when the possibilities for soil volume increase are limited, dilatancy can cause changes in the stress condition during shear.

1.6 EXAMPLES OF FREE AND CONSTRAINED DILATANCY

Let us place a lengthy band, undergoing a normal pressure σ_{n_o}, on the surface of dense non-cohesive soil (Fig. 1.8a). The application of shearing force causes the mobilization of resistance to contact shear under the band (Fig. 1.8b), which reaches the ultimate value

$$\tau_u = \sigma_{n_o} \, \text{tg} \, \varphi_c \tag{1.6}$$

where φ_c is the angle of contact friction.

The mobilization of resistance to shear is accompanied by the failure of initial grain packing in a layer of certain thickness. It is within this layer that we can observe repacking with loosening of soil grains to the critical density – that is, dilatancy. But while determining resistance to shear it is not necessary to allow for dilatancy, because the stress condition remains unchanged all throughout the process ($\sigma_n = \sigma_{n_o}$ = const). What does change is the angle of shear, which can reach the maximum value φ_c. In this latter the indirect influence of dilatancy upon the value is taken into account.

An analogous example would be true for non-cohesive soil, where shear is accompanied by contraction with compaction of the contact layer. Here the change of density in the process of failure does not influence the value of the acting normal pressure, either.

Now we shall turn to the mechanism of constrained dilatancy.

Let us place the band under consideration into the same dense grainy soil at a

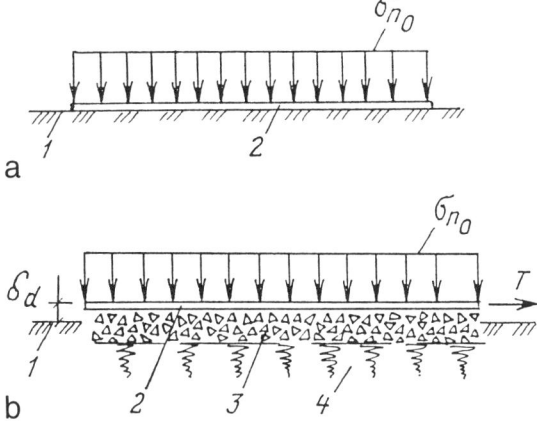

Figure 1.8. Mechanism of free dilatancy-shear of a plate along the soil surface: a) initial stress condition; b) condition after overcoming the forces of resistance to shear. 1. soil surface; 2. sheared plate; 3. grain layer which undergoes repacking during the shear (dilating layer); 4. massif of soil whose grains do not undergo mutual repacking.

certain depth corresponding to normal pressure generated by the soil's own weight $\sigma_n = \sigma_{n_o}$ (Fig. 1.9a).

The application of the tearing-out load to the band leads to mobilization of contact resistance to shear, and formation of a layer of failure of initial grain packing. Dilatancy occurs (Fig. 1.9b). But the development of dilatancy requires the wringing of the soil massif adjacent to the contact layer. Such wringing must cause the development of additional stresses of dilatant thrust $\Delta\sigma_d$. These stresses reach their maximum in the condition of limiting equilibrium, when the band starts to slide against the soil with a simultaneous reaching of critical density in the contact layer.

As a result, $\sigma_n \neq \sigma_{n_o}$, and the ultimate resistance to shear will determine the aggregate normal pressure $\sigma_n = \sigma_{n_o} + \sigma_d$, i.e. when the sheared body is constrained, the stress condition does not remain constant. Dilatancy, when it is constrained, becomes one of the factors determining the resistance of grainy medium to shear.

A similar example can be given also for the case when the sheared body is constrained in loose non-cohesive soil. Here the contraction emerging in the process of failure is displayed in the drop of initial normal pressure, and in the condition of limiting equilibrium its value is determined by the difference $\sigma_n = \sigma_{n_o} - \sigma_d$.

The examples given above prove that there exists a disparity between soil failure in the conditions of free and constrained dilatancy. In the latter case the condition of shear and soil deformation active outside the failure zone. The stress condition of failure zone is determined by the thrust of the adjacent massif, which is evidently a function of the massif's deformational characteristics. The above-

18 *Dilatancy as a fundamental property of granular media*

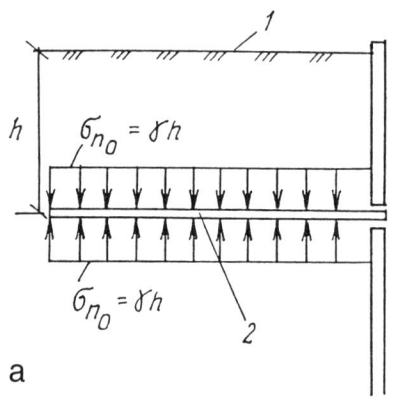

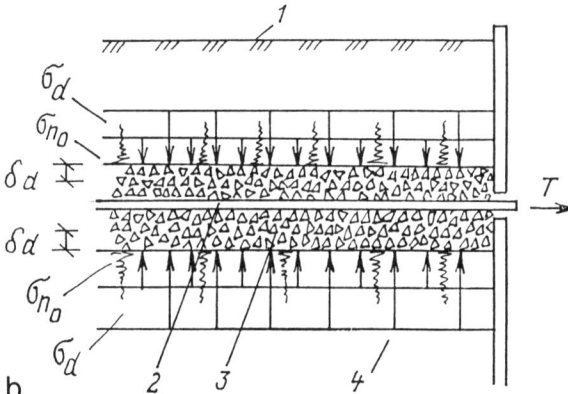

Figure 1.9. Mechanism of constrained dilatancy – shear of a plate within the soil massif:
a) initial stress condition;
b) stress condition during the shear – pressure developed as a consequence of the dilatant thrust σ_d is added to the initial normal pressure. 1-4: See Fig. 1.8.

mentioned correlation between strength and deformational characteristics comes into force. This is especially important while choosing a method of determining soil strength parameters. The frequently observed discrepancy between the calculated and real soil strength appears to be the consequence of ignoring the dilatancy factor.

1.7 MANIFESTATION OF DILATANCY IN SOIL STRENGTH TESTS

The existing notions of soil strength are inseparable from the methods of determination. In the majority of conducted laboratory tests direct shear instruments are used, stabilometers and, rather infrequently, instruments with three separate loading platforms. Let us turn now to the analysis of failure conditions modelled in these instruments.

In direct shear instruments (the most widely used modification) the test aims to determine the ratio between normal and shearing stress. The application of normal pressure σ_n is effected as a rule by means of a system ensuring free vertical deformation of the dilating sample. Thus throughout the process of loading

$\sigma_n = \sigma_{n_o}$ = const, and the test de facto ensures the measurement of ratio between the shearing and initial normal pressure which can maximally reach the value tg φ_o. In this way the conditions of free dilatancy of soil are modelled during the test. The obtained chart of the function $\tau_u = f(\sigma_{n_o})$ is represented in Fig. 1.3.

If we preclude the vertical advance of the piston the soil sample will manifest absolutely different resistance to shear. In the case of dense soil it will turn out considerably higher, and the chart $\tau_u = f(\sigma_{n_o})$ will manifest the value of the angle $\varphi_1 > \varphi_o$.

It may happen that the value φ_1 will exceed 45°. Rather often similar measurements of $\varphi > 45°$ lead to errors in conclusions. The ratio between the shearing and initial normal pressure is treated as the angle of internal friction which has received an increment as a result of dilatancy. It was convincingly demonstrated by Wernick (1977) in his research dealing with the mechanism of sand shear along the lateral surface of an injection anchor root. According to Wernick (and a number of other researchers) in dilating soils there is a violation of the condition of coaxiality between the increments of stresses and deformations. In this case an allowance for the so-called angle of dilatancy $\Delta\varphi_d$ must be made in calculating the angle of internal friction, i.e.

$$\varphi' = \varphi_o + \varphi_d \qquad (1.7)$$

where φ_o is the angle of internal friction of soil determined in the course of a standard test.

The maximum value of φ' measured by Wernick in the apparatus of the so-called 'true direct shear' constituted 53.1° for coarse sand.

While interpreting similar tests one should keep in mind that dilatancy in the process of shear causes the development of thrust stresses. As a result of this normal pressure increases by the value of dilatant pressure (σ_d), and the value of the angle of internal friction will determine the ratio between ultimate shearing and final normal stress, i.e.

$$\varphi' = \arctan \frac{\sigma_{n_o} + \sigma_d}{\tau_u} \qquad (1.8)$$

It is understood that the increment of pressure caused by dilatancy is in direct proportion to the rigidity of the connection which limits the increase of the sample volume. This fact was emphasized by Dranovsky & Rossikhin (1983); Dranovsky (1985); Dranovsky & Vorobiov (1985, 1986), who conducted tests in a direct shear apparatus where the application of normal pressure was effected by an elastic link (dynamometer). They showed that elastic limitation of the piston's upward travel leads to a considerable increase of normal pressures and growing resistance to shear.

Constraint of dilatancy may also cause errors in standard shear tests due to the friction of soil against the walls of the sample ring. This phenomenon was studied by Amsheyus & Kuleshus (1982) on a direct shear apparatus, in the lower movable carriage of which pressure detectors were installed. Amsheyus demonstrated that normal stresses in the shear plane increased by 30% or more for coarse sands, whereas for loose sands these stresses were considerably lower.

Strong friction against the walls of the sample ring (especially in the case of coarse sands) does not allow to measure precisely the value of real normal pressure in the shear plane. Such constraint of dilatancy results in the appearance of the so-called 'gear cohesion' in the chart $\tau_u = f(\sigma_{n_o})$ (Fig. 1.10). Goldstein (1979) points out, that the denser is the sand – the higher is resistance to initial wedging out of grains (dilatancy), the wider the shear zone should be, the higher the relative importance of gear in resistance to shear becomes.

Similar to shear tests, standard tests in triaxial apparatus also presuppose the condition of unrestrained volume change. In sample 'squash' tests the existing loading systems ensure constant skin pressure and allow the liquid from the instrument camera to flow freely into the volumometer. In this case only the increment of volume in the process of failure can be measured. On the other hand, when the flow of liquid is constrained, the skin pressure increases at the expense of dilatant thrust, and due to this the limiting ratio between principal pressures changes.

During triaxial tests the sample loading method takes account of the parameter reflecting the ratio between principal stresses and is expressed as follows:

$$\mu_\sigma = \frac{2\sigma_2 - \sigma_1 - \sigma_3}{\sigma_1 - \sigma_3} \qquad (1.9)$$

A modification of test conditions can considerably change the measured strength parameters. This is illustrated by charts in Fig. 1.11a, b borrowed from Malyshev (1980). Thus, according to Lomize & Crighanovski the value of φ for the same sand can fluctuate from 35° in 'squash' tests ($\mu_\sigma = -1$) and 'tension' tests ($\mu_\sigma = +1$) to 48° in torsion tests ($\mu_\sigma = 0$). As it is underscored by Malyshev (1980), the less the absolute value of the angle of internal friction is, the less is the

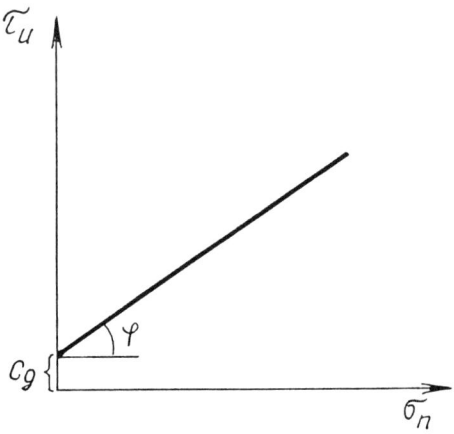

Figure 1.10. Cohesion of gear with shear of coarse sand (Goldstein, 1979).

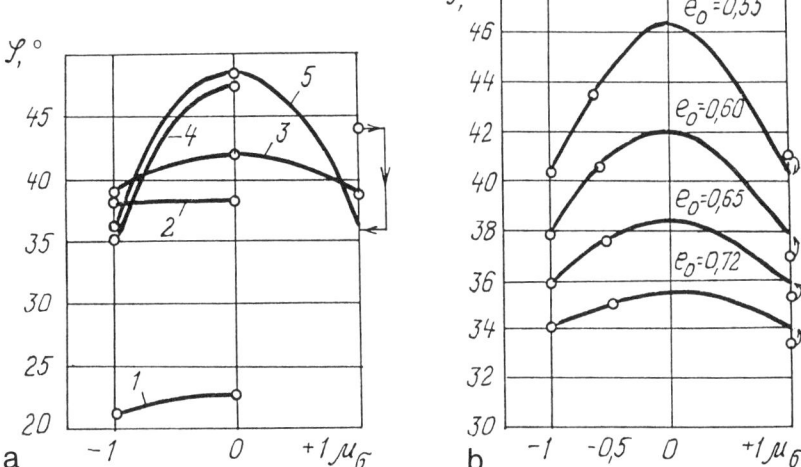

Figure 1.11. a) Dependance of the angle of internal friction (according to Mohr) on the parameter μ_σ (according to various authors) (Malyshev, 1980): 1. Malyshev; 2. Barshevsky; 3. Kirkpatrick; 4. Stroganov, Lomize & Krighanovski; b) the, same according to Cornforce with different values of initial coefficient of porosity.

influence of μ_σ on its value. Evidently, this is due to the fact that the decrease of μ_σ indirectly signifies a decrease of dilatancy.

Modifying the test conditions always means modifying the conditions of deformation of the sample. These latter are programmed by the medium into which the sample is placed, and by the reaction of this medium to dilatancy in the process of failure. This can be proved by the functional relationship of the Poisson coefficient ν to the dilatant/contractional changes in the process of shear. As Goldstein (1979) points out, dilatancy may and does cause the measurement of values of ν > 0.5 for extra-dense soils, whereas contraction of loose soil sometimes causes the values of ν ≤ 0 in the process of monoaxial compression. In this light it seems justified to speak of the influence of medium deformability on strength, both when the medium surrounds the plastic deformation zones or is adjacent to them. Such influence can be allowed for by means of a method which enables the researcher to find the functional relation between strength parameters and deformational characteristics of the soil medium.

1.8 CONCLUSION

Traditional soil testing methods are oriented on free dilatancy conditions. When these conditions are violated this invariably leads to distortions and contradictions in interpreting experimental data, especially when these data are compared with in-situ tests. These contradictions have put on the agenda the problem of updating the existing soil strength theories. At the same time no due attention is paid to dilatancy as a major factor determining the mechanism of grainy medium failure.

We should also keep in mind the fact that the existing strength theories are designed to be applied to solid mediums which do not possess the property of changing their volume in the process of distortion. If, as Reynolds (1985) justly remarked, dilatancy is a fundamental property which is not characteristic for certain liquids and solid bodies, then, apparently, strength theories have to be re-read allowing for this property. This is possible if we introduce into soil mechanics phenomenological models reflecting the dilatancy factor. It seems appropriate here to mention the words of Nikolaevsky (1975): major perspectives of soil mechanics development are connected with qualitative research of the main objectives in the framework of elasto-plastic dilatantial models.

CHAPTER 2

Models of elasto-plastic deformations of dilating non-cohesive soil

2.1 GENERAL PROPOSITIONS

The phenomenological models of soil behaviour in the process of shear and internal bulge set forth underneath have been put into the foundation of special instruments, test methods and calculation schemes. Each of these models is a combination of a sliding model and an elastic model. This combination is construed on the division of plastic deformations in shear zones and elastic deformations in adjacent areas. The models allow for the influence of constrained dilatancy on the stress condition and soil failure conditions. A specific instance is the lack of constraint on dilatancy, when such influence is not observed.

A number of concepts have been introduced to construe the models:
– Layer or band of shear, where the dilatancy – accompanied deformations of sliding or grain-repacking are localized;
– Areas or zones of failure for instance when sliding deformations spread over a certain volume;
– Massif, where no grain-repacking occurs, and elastic interaction mainly takes place at contact points instead.

Dilatancy is considered to be the result of both purely plastic and partially elasto-plastic deformations localized in the shear layer or failure volume. Outside these latter, within the massif, deformations connected with grain sliding are completely absent. Thus, the deformation of massif is programmed by the value of dilatancy, whereas its reaction is conditioned by the values of elastic characteristics of soil with intact structure. Strictly speaking, near the shear zone soil can experience local deformations of compaction which are also plastic. Therefore in order to avoid possible contradictions we shall assume that in further chapters dilatancy is understood as a movement of a conventional border between the shear layer or failure zone and the massif, i.e. elastic compression or unloading (in the process of contraction) of the massif.

2.2 MODELS OF CONTACT SHEAR

Two propositions are included in the basis of shear models:
1. Soil failure in the process of shear is localized in the layer (band) along the contact surface;
2. Dilatancy of the shear layer causes elastic resistance of the soil massif.

This presupposes that purely plastic (deviatory) deformation takes place in the shear layer, whereas elastic (tensor) deformation occurs in the massif.

To describe the massif deformation we shall make use of the modulus of elasticity E, and the coefficient of elastic resistance (coefficient of uniform compression)

$$K = \frac{\Delta\sigma_d}{\Delta\delta_d} \qquad (2.1)$$

which reflects the proportional relationship between the increment of dilatant stress $\Delta\sigma_d$ and the corresponding movement $\Delta\delta_d$. These quantities are connected by the following expressions:
 – for an axis-symmetric sum

$$E = (1 + \nu)rK, \text{ and} \qquad (2.3)$$

 – for a plane sum

$$E = (1 - \nu^2)\omega bK \qquad (2.3)$$

where r, b are, respectively, radius and width of the shear surface; ν is Poisson's coefficient; ω is coefficient of the form of sheared surface (Tsytovich, 1963).

We assumed here that the coefficient ω corresponds to mean compression (loading) of soil massif as a result of dilatancy; the compression is perpendicular to the shear surface. The values of ω at different values of η (ratio of the larger (l) and smaller (b) sides of the sheared surface) are listed in Table A1.1.

Dilatant movement $\Delta\delta_d$ accompanying the compression or unloading of massif is assumed here to be either the broadening or the narrowing of the shear band. Its direction is normal relative to the conventional border between the layer of plastic grain sliding deformations and the elastic deformation zone in the massif, and this direction coincides with the direction of dilatant stress $\Delta\sigma_d$.

Phenomenological model of contact shear in conditions of constrained dilatancy is represented in Fig. 2.1. Fig. 2.1a reflects the initial stress condition of soil, when a certain initial normal pressure σ_{n_o} is active on the contact surface. This pressure is transmitted through the grain layer of soil by means of elastic springs which model the massif.

Deformability of springs is characterized by their rigidity S which reflects the ratio between the increment of compression force ΔP and the corresponding movement. In the case when compression is caused by the dilatancy of contact layer grains, the rigidity of springs can be expressed as follows:

$$S = \Delta P_d / \Delta \delta_d \tag{2.4}$$

where ΔP_d is the increment of force due to dilatancy. When the surface is sheared by the area A

$$\Delta P_d = \Delta \sigma_d A \tag{2.5}$$

then taking into account (2.1)

$$S = \Delta \sigma_d / \Delta \delta_d A = KA \tag{2.6}$$

In the model under study the spring length corresponds to the thickness of massif zone within which the additional stresses emerging in the process of dilatancy are distributed. This thickness is in direct proportion to the area (radius) of the shear surface i.e. proceeding from (2.2) and (2.3), one and the same dilatant movement $\Delta \delta_d$ can cause different increments of normal pressure $\Delta \sigma_d$ in the massif.

Application of the shearing force causes dilatancy due to the mutual wedging out of grains within the contact layer (Fig. 2.1b). As for the springs modelling the reaction of massif, additional compression emerges, which leads to the increase of pressure on the shear surface. This pressure reaches its maximum in the limiting equilibrium state, when it equals

$$\sigma_n = \sigma_{n_o} + \sigma_d \tag{2.7}$$

If we take into account, for example, (2.1), (2.2), then

$$\sigma_d = \frac{E}{(1+\nu)r} \Delta \delta_d \tag{2.8}$$

i.e. $\sigma_d = f(K, E)$, and, consequently, $\sigma_n = f(K, E)$, (2.9)

i.e. the reaction of the massif to dilatancy or its deformability dictates the conditions of contact layer failure, which happens when critical density of grain packing in the massif is achieved. Keeping in mind that, according to Coulomb's law (1.3), ultimate resistance of non-cohesive soil to shear τ_u is in direct proportion to active normal pressure, it would be logical to suggest that for the case of constrained dilatancy

$$\tau_u = f(K, E, \varphi) \tag{2.10}$$

Zero spring rigidity according to the suggested model is a particular condition for the case of free dilatancy. Then, actually

$$\tau_u = f(\varphi) \tag{2.11}$$

If we reverse Fig 2.1a and 2.1b, we get the picture of the contraction mechanism – when there is a decrease of the contact layer thickness in the process of shear. In this case normal pressure in the process of failure is determined by the difference of initial and dilatant pressures, i.e.

26 *Models of elasto-plastic deformations of dilating non-cohesive soil*

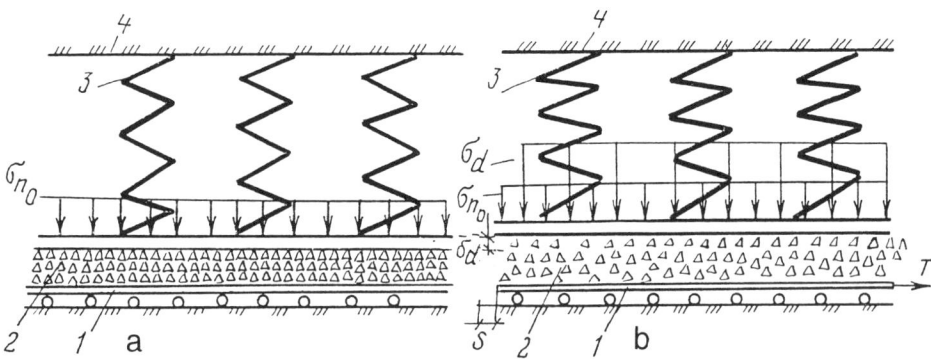

Figure 2.1. Constrained dilatancy contact shear model: a) initial stress condition – normal pressure σ_{n_o}; b) limiting equilibrium condition – normal pressure is equal to $\sigma_{n_o} + \sigma_d$. 1. contact surface; 2. sheared (dilating) grain layer; 3. springs modelling elastic resistance to dilatancy on the side of the massif; 4. border of the zone wherein additional dilatantial stresses are distributed.

$$\sigma_n = \sigma_{n_o} - \sigma_d \qquad (2.12)$$

Other relationships considered above remain valid.

The state when an infinitely small increment of the shearing load causes rolling of contact layer grains, corresponds to limiting equilibrium. It is the case of the so-called kinematic equilibrium, which is characterized by constant sliding speed and constant value of shearing stress τ_u^k. A certain lowering of the load (less than τ_u^k) results in fading out of the movement due to the recovery of the grain gear.

The model in Fig. 2.1 reflects contact shear along the surface of a rigid element which does not undergo considerable interior deformations when loaded. This can be applied to large-diameter piles or trench foundations (diaphragm walls). Interior deformations of the sheared body should not be ignored, though, for instance in the case of micropiles and injection anchor roots (they experience lateral compression or thrust, when loaded see Chapter 11), and also reinforcement elements in soil (they undergo considerable stretching).

The influence of lateral deformations on contact shear models can be represented by way of placing springs under the sliding band (the spring's rigidity must conform with deformability of the sheared body; the springs can also be made of lead if the sheared body experiences plastic deformations, (see Fig. 2.2). In the process of the dilatant thrust build-up these springs will be compressed, modelling the influence of the decrease of cross section on the value of dilatancy.

Shear conditions when the sheared body is stretched lengthwise are represented in Fig. 2.3 in the example of a reinforcement element squeezed on both sides in a dilating massif. The stretching of the reinforcement element mobilizes resistance to shear along contact surfaces (Fig. 2.3b), but it is also capable of considerably dampening the dilatant thrust suppressing due to the grains pulling apart, i.e. due to the gradual decrease of density in the contact layer. This example

'Soil-to-soil' shear model 27

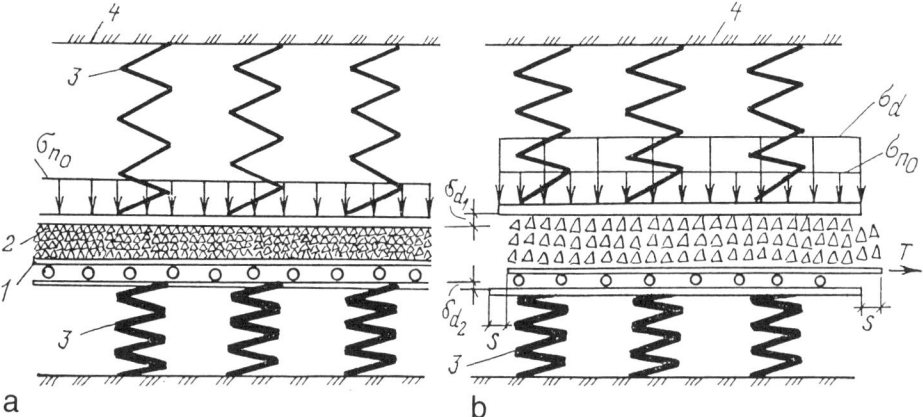

Figure 2.2. Model of shear with lateral deformation of the contact surface: a) initial stress condition; b) limiting equilibrium condition. 1. contact surface; 2. sheared (dilating) grain layer; 3. springs modelling from the top: resistance to dilatancy on the side of the massif; from the bottom: lateral deformation of the sheared surface; 4. border of the zone wherein additional dilatantial stresses are distributed.

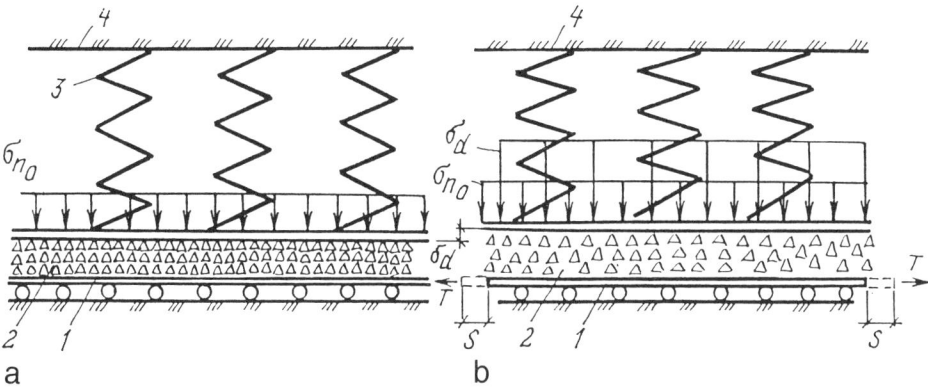

Figure 2.3. Model of shear with lengthwise stretching of the contact surface: a) initial stress condition; b) limiting equilibrium condition. 1–4. see Fig. 2.1.

allows us to evaluate the behaviour of reinforced soil as a composite material, for which the process of deformative development under the load and the mechanism of failure are determined by the ratio of deformative and strength parameters of reinforcement and fill (see Chapter 12).

2.3 'SOIL-TO-SOIL' SHEAR MODEL

Such a model is represented in Fig. 2.4. It differs from contact shear models by the arrangement of springs modelling the plastic resistance to dilatancy on both sides of the shear band. The aggregate rigidity of massif divided by the line of sliding

28 *Models of elasto-plastic deformations of dilating non-cohesive soil*

determines the increment of normal pressure in the process of shear. A model reflecting the mechanism of non-cohesive soil contraction can be obtained by reversing Figs 2.4a and 2.4b.

The condition, in which elastic characteristics of massif on both sides of the dilating layer of sliding surface are different, can be modelled by installing springs with different rigidity. The aggregate massif rigidity in this case will be characterized by the value of modulus of elasticity (see above) and coefficient of elastic resistance.

Let us consider the deduction of the value of modulus of elasticity given above. According to the model in Fig. 2.4, the increment of stress due to dilatancy $\Delta\sigma_d$ has the same effect on both sides of the massif divided by the shear band. But with different values of their rigidity, one and the same elastic reaction to dilatancy results in different values of movement.

Let us assume that on one side of the shear band the massif has a modulus of elasticity with the value E_1, and a coefficient of elastic resistance with the value K_1, and on the other side of the shear band – respectively E_2 and K_2.

Then the above-mentioned modulus of elasticity will be determined by the ratio

$$E_{\text{red}} = \frac{\Delta\sigma_d}{\varepsilon_{d_1} + \varepsilon_{d_2}} \qquad (2.13)$$

where $\varepsilon_{d_1} + \varepsilon_{d_2}$ is the aggregate relative dilatant deformation.

If we take into consideration, that

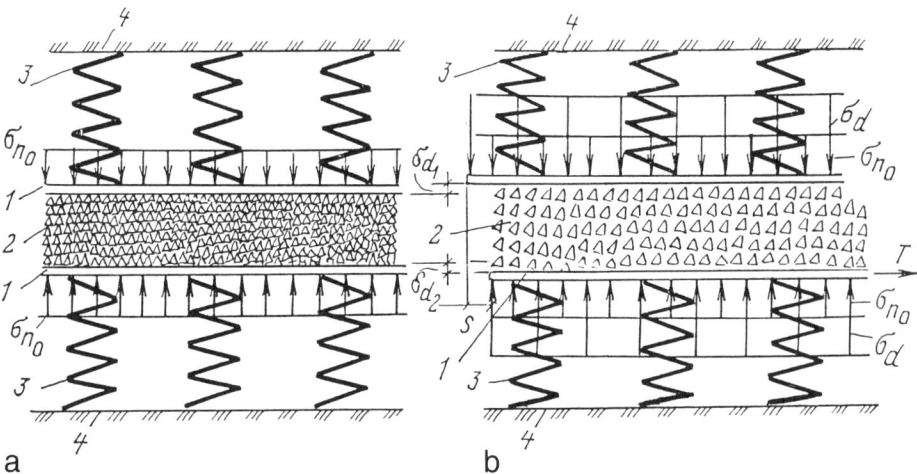

Figure 2.4. 'Soil-to soil' shear model: a) initial stress condition – normal pressure is equal to σ_{n_o}; b) limiting equilibrium condition – normal pressure is equal to $\sigma_{n_o} + \sigma_d$. 1. borders of grain-repacking zone (zone of dilatancy); 2. shear band; 3. springs modelling elastic resistance to dilatancy on both sides of the massif; 4. border of the zone wherein additional dilatantial stresses are distributed.

$$\varepsilon_{d_1} = \frac{\Delta\sigma_d}{E_1} \text{ and, } \quad \varepsilon_{d_2} = \frac{\Delta\sigma_d}{E_2} \tag{2.14}$$

then

$$E_{\text{red}} = \frac{E_1 E_2}{E_1 + E_2} \tag{2.15}$$

Similarly, it is possible to deduce an expression for the above-mentioned value of coefficient of elastic resistance:

$$K_{\text{red}} = \frac{K_1 K_2}{K_1 + K_2} \tag{2.16}$$

Where $\Delta\delta_{d_1} + \Delta\delta_{d_2}$ is the aggregate dilatant compression (stretching) of the massifs.

Thus, for dilating along the surface the relationship between E and K can be expressed as

$$E_{\text{red}} = (1 - v^2)\omega b K_{\text{red}} \tag{2.17}$$

when it is necessary to allow for the difference of Poisson's coefficients, it is possible to adopt the mean value.

2.4 MODEL OF INTERNAL BULGE

Soil deformation conditions in the process of omnilateral compression are reflected by the model represented in Fig. 2.5. The area of plastic deformations is

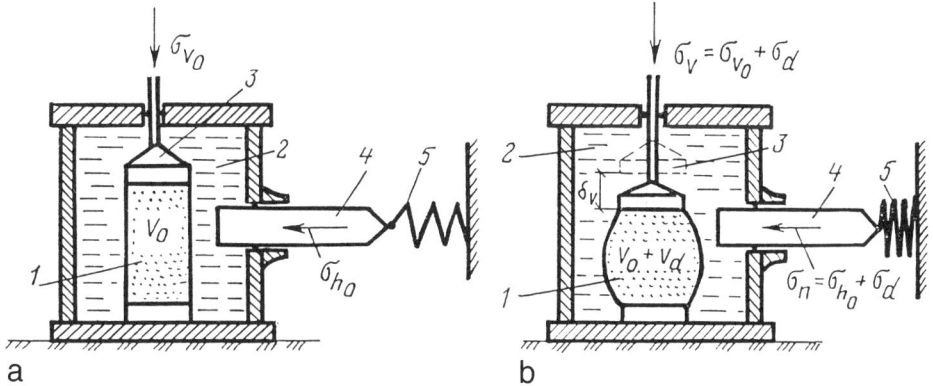

Figure 2.5. Model of internal thrust of soil: a) initial stress condition – initial vertical pressure σ_{v_o} and initial horizontal pressure σ_{h_o} are active; b) maximum limiting condition – the mobilized vertical and horizontal pressures have received an increment by the dilatant $+\sigma_d$. 1. area of internal thrust; 2. liquid in a hermetic camera; 3-4. pistons of vertical and horizontal pressure; 5. spring, modelling the elastic reaction of massif to dilatancy.

presented as an impermeable shell with soil having a certain initial density of packing. The shell is placed inside a closed liquid-filled camera. One of the camera's walls is fitted with a cartridge consisting of a piston and a spring of certain rigidity. This cartridge allows to create initial pressure σ_{h_o} on the walls of the shell and to model the reaction of massif to volume deformations of soil (Fig. 2.5a). When the shell with soil is loaded with the help of the upper piston, there is a change of density of grain packing which is accompanied by volume dilatancy. This latter is taken up by the liquid within the camera and transmitted by means of a piston) to the spring, whose rigidity models the elastic reaction of the massif. The ultimate pressure on soil will correspond to the end of volume deformations, i.e. to the moment when critical density of grain-packing is achieved. Simultaneously, the maximum value of dilatant thrust is reached as well (Fig. 2.5b).

The stress condition of soil in the failure area is characterized in this model by normal stress components: vertical σ_v and horizontal σ_h. When the former increases, the initial horizontal stress σ_{h_o} receives an increment of the value of the dilatant $\Delta\sigma_d$, which is determined by the spring rigidity or by the volume of liquid which it can displace. Thus, in this case dilatancy should be regarded as a volume deformation. But if we examine the break-down of movements accompanying it according to the directions along which principal normal stresses are active, it appears possible to calculate the changes in the stress condition as increments of principal normal stresses by the value determined by ratios of the (2.8) type.

The ultimate ratio between principal stresses (characterized by Rankine formula (1.4)) corresponds to the maximum value of increment of horizontal stress due to dilatancy

$$\sigma_h = \sigma_{h_o} + \sigma_d \tag{2.18}$$

While reading the Rankine formula, though, one should keep in mind that in the case of constrained dilatancy both the components of principal normal stresses are a function of deformative characteristics of the massif, i.e.

$$\sigma_{1,3} = f(E, K, \varphi) \tag{2.19}$$

The suggested model also allows to represent the case of unloading the area of plastic deformations. For this purpose it is necessary to lower the pressure on the upper piston (Fig. 2.5). The changing stress condition will correspond to the new value of critical density and will also lead to the volume deformation of soil. In accordance with the spring rigidity, the unloading of the horizontal stress in the massif will be taking place, i.e.

$$\sigma_h = \sigma_{h_o} - \Delta\sigma_d \tag{2.20}$$

2.5 JUSTIFICATION OF THE ASSUMPTION REGARDING ELASTIC REACTION OF THE MASSIF TO DILATANCY

The assumption regarding elastic behaviour of the massif presupposes the use for describing the model of the modulus of elasticity E, a quantity rather seldom employed in soil mechanics. Traditionally, the so-called modulus of deformation is used, which reflects the proportionality between the increment of normal pressure and a certain fixed settlement. In the process of calculating the modulus of deformation during stamp and compression tests, and in stabilometers, soil experiences both purely elastic reversible deformations and plastic deformations of compacting, and the task is to determine the acceptable share of the latter. Due to this fact it is unjustified to use the modulus of deformation as a quality, characterizing the massif's behaviour in its reaction to dilatancy.

Let us return to the analysis of the character of soil deformations at the level of interaction between separate grains. Plastic deformation of non-cohesive soil can be connected only with the appearance of sliding at grain-contact points and the grains' breaking. Both these phenomena are irreversible and cause changes in the initial grain packing pattern. On the contrary, reversible deformations are determined only by deforming grains at contact points without changing their original packing. Such deformation can signify compression of the grain material proper, and of bound-water films.

Apparently, purely elastic behaviour of the grainy medium is possible only when the values of movements are very insignificant, and when these movements are caused by some exterior action. For sands these values are in the order of fractions of a millimetre, i.e. here we are dealing with values seldom, if at all, taken into account in traditional geotechnical practice.

But it is such small movements that transpire in the process of dilatancy. For instance, in the conditions of free expanding of the sample during shear-instrument tests the values of vertical movement rarely exceeds one millimetre even for coarse sands. In the case of finer sands, or when dilatancy is constrained, these values go down to tenth and hundredth fractions of a millimetre. Similar movements are not capable of causing grain repacking or any considerable plastic deformations in the course of compression or unloading of the massif adjacent to the shear layer.

It is rather difficult to recommend specific methods of calculating E, which can be accounted for by these methods' not being worked out well enough. It appears that the use of geophysical methods of modulus calculation by tracking of longitudinal waves is preferable (Erykhov, 1961; Hope 1989). As a first approximation it is possible to use punch or compression tests with multiple repetition of loading and unloading (sectional modulus). Rough values of E calculated by this method are given by Goldstein (1979) and borrowed by us for Table 2.1.

Table 2.1 leads us to the conclusion that there is a tendency for the modulus to increase parallel to the increase of density of soil, grain size and grain roughness.

32 *Models of elasto-plastic deformations of dilating non-cohesive soil*

Table 2.1. Values of sectional modulus of elasticity E after multiple repetition of loading.

Type and form of sand grains	Value of E, MPa	
	Loose	Dense
Quartz grains, fine angular	120	210
Quartz grains, medium angular	135	190
Fine grains, rolled	180	–
Medium grains, rolled	210	320
Medium grains, slightly angular	140	240
Medium grains, slightly angular but well sorted-out	100	200

The last fact is possibly connected with a less stable equilibrium at contact points for angular grains, and the fact that they are more prone to break when loaded.

For practical calculations we recommend the formula suggested by Barkan et al. (1974), according to which

$$E \approx 8E_o, \qquad (2.21)$$

where E_o is the modulus of aggregate sand deformation.

2.6 CONCLUSION

Consideration of possibilities given by the models described above allows to draw a number of preliminary conclusions regarding the factors determining the influence of dilatancy on the stress condition and failure processes. Thus, we can note that the increment of normal pressure is in direct proportion to the size of grains in the dilating layer, the density of their initial packing and the rigidity of the massif. A decrease of density lowers the thrust appearing during shear, while a comparatively loose grain packing can lead to compaction which is dangerous due to a possible rapid failure when normal pressures drop. Evidently, there is a certain grain size when dilatancy becomes a minor phenomenon due to the increased surface activity of grains. In this case the term 'particle' should be used instead of 'grain'. An exact differentiation of these two terms is essential for a correct reflection of the qualitative difference between the respective sets of phenomena.

The presented models also supply an explanation of the mechanism of a temporary drop of porous pressure in water-saturated sands when a dilatancy-accompanied shear takes place, and a rise of this pressure when contraction occurs. These phenomena described by Reynolds (1886) can be observed in non-drained shear tests. Another interesting phenomenon is the rise of porous pressure in shear zones when contraction of loose fine sands takes place; it lies in the basis of liquefaction which is described in Section 12.2.

The suggested models do not exclude the possibility to take into consideration the reology of soils. For example, if we introduce into our calculations the time necessary for fading out of dilatancy-connected movements in the grainy medium, it will probable enable us to allow for the phenomena of relaxation of dilatant stresses and provide a theoretical explanation of the decrease of resistance to shear, to calculate short-term and long-term soil strength. It can be practically tested in the suggested models by, for example, lowering the spring's force gradually during the course of the experiment. As for the sums dealing with the migration of pores' water, its influence on the general level of stresses has to be specially allowed for, while designing instruments and test methods.

CHAPTER 3

Instruments and methods for soil testing in the conditions of constrained dilatancy

3.1 GENERAL PROPOSITIONS

This chapter dwells on the construction of special instruments designed for shear and triaxial tests. Phenomenological models for the failure of dilating soils described in chapter 2 have been realized in these instruments. All the presented instruments allow to conduct experiments in conditions of both free and constrained dilatancy. In the process of designing an attempt was made to widen the range of their possibilities and avoid errors which may appear due to an inadequate reflection of natural conditions of soil deformation and failure. Parallel to this methods of conducting tests were given. Some of these methods were practically employed by the author, while others are proposed for further research.

3.2 DILATOMETRIC APPARATUS OF CONTACT SHEAR (DACS)

This instrument is defended by a USSR copyright, certificate 1491143 under the name 'Apparatus for measuring contact resistance to shear of dry materials' (Sobolevsky & Popov, 1989). Its design corresponds to the model of contact shear described in Section 2.2 (Figs 2.1 and 3.1). Two modifications of the instrument were manufactured, namely: DACS-1 and DACS-2. The former to conduct tests with one-time application of the shear force, whereas the latter ensures the possibility of cyclic testing. The design of DACS-1 is represented in Fig. 3.2. The instrument is mounted on a slab (1) holding a rest (2) for applying the shear force from a screw jack (3) through a dynamometer (4); movable lower carriage (5) on bearings (6); posts (7) for fixing an upper carriage (8) and a box (9) with a soil sample (10), frame (11) for a screw jack of normal pressure (12) and regulated dynamometer (13), clock-type indicator (14) with the accuracy of 0.01 mm which is fixed on a conical piston (15) for decreasing the friction against the wall of the box (9).

The soil sample is placed into a square-section box with an area of 100 cm^2.

Figure 3.1. Dilatometric apparatuses of contact shear: a) DACS-1; b) DACS-2.

The clearance between the box (9) and carriage (5) is regulated by screws (26). The movement of the carriage (5) is controlled by the clock-type indicator (16) with the accuracy of 0.01 mm.

The DACS-2 instrument is distinguished by a symmetric arrangement of rests (2) and dynamometers (4) on both sides of the carriage (5). When water-saturated sands are tested the lower carriage (5) is furnished with a plastic tub.

The design of the movable carriage (5) includes a lift-off plate (17) with corrugated surface used for gluing on the tested soil grains, geotextile, applying layers of cement mortar or other materials for which the contact friction is measured. Under the lift-off plate (17) in the carriage (5) there is a chamber for installing a cell for direct measurement of normal pressure in the shear plane. The

36 Soil testing in the conditions of constrained dilatancy

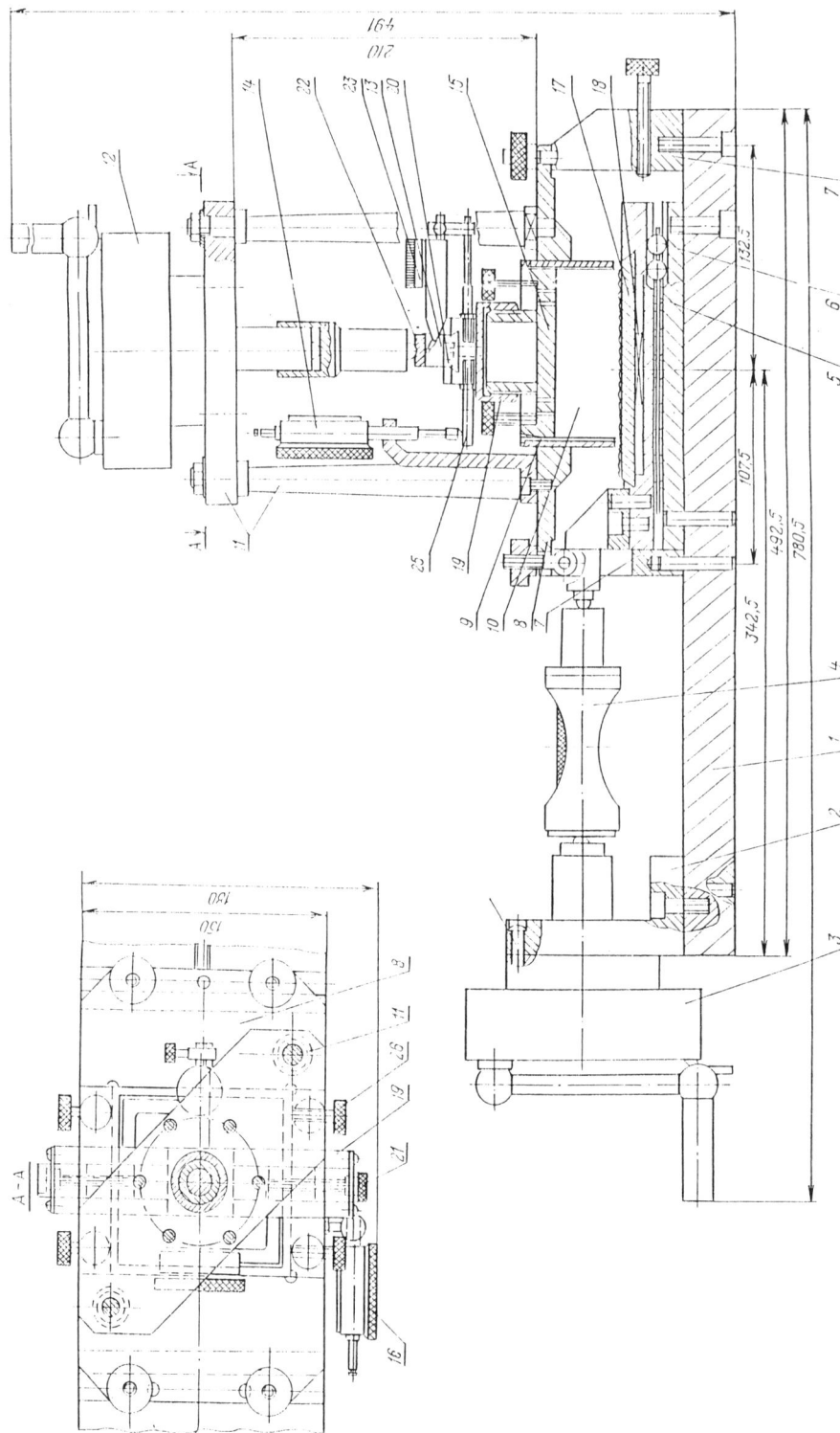

Figure 3.2. DACS-1 design. 1. slab; 2. rest; 3. jack; 4. dynamometer; 5. lower carriage; 6. bearings; 7. post; 8. upper carriage; 9. sample box; 10. sample; 11. frame; 12. jack; 13. regulated dynamometer; 14. indicator; 15. piston; 16. indicator; 17. lift-off plate; 18. pressure cell; 19. scale of dynamometer (13); 20. wedge-formed supports; 21. regulating screw; 22. elastic beam; 23. beam indicator; 24. dynamometer installation cartridge.

value of ultimate movement (compression) of the cell must be minimal (not exceeding the order of 10^{-3} mm) to avoid its possible influence on the stress condition of the sample.

The key aggregate part of the instrument is a regulated dynamometer (13) (Fig. 3.3). The dynamometer consists of a scale (19) with movable wedge-shaped supports (20) regulated by screws (21); an elastic beam (22) whose deflection is measured by indicator (23) with an accuracy of 0.01 mm. Cartridge (24) serves to fix the dynamometer on the piston, and the grip (25) – to mount the indicator (14) which registers the dilation (compaction) of the sample due to the dilatancy in the process of shear. The force is applied by the screw jack (12) in the middle of the beam span.

The regulated dynamometer (13) is calibrated beforehand at different values of span between supports (20). During the calibration the dynamometer's rigidity is

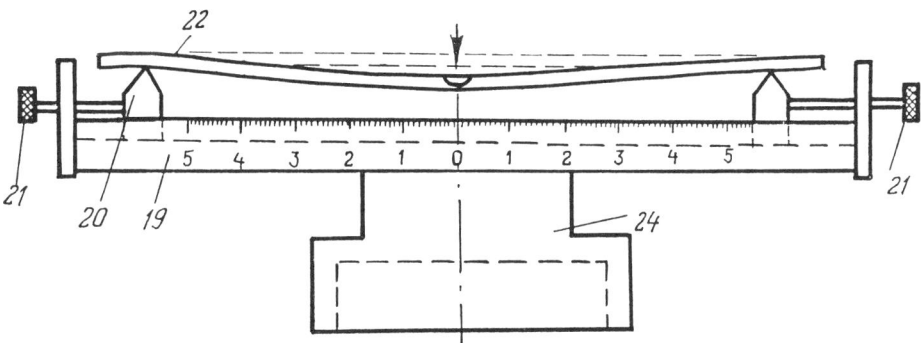

Figure 3.3. Regulated dynamometer's design: 19. scale; 20. wedge shaped supports; 21. regulating screw; 22. elastic beam; 24. dynamometer installation cartridge (force applied by the jack).

calculated as

$$S = \frac{\Delta P}{\Delta \delta} \qquad (3.1)$$

It reflects the ratio between the increment of force ΔP and the deflection of the beam $\Delta \delta$. As for the rigidity of the beam, it is related to the elastic resistance coefficient (coefficient of uniform compression) by the ratio (2.6).

Returning to Section 2.2, let it be remembered that the coefficient of elastic resistance K for soil is expressed by the ratio between the increment of dilatant stress $\Delta \sigma_d$ and the movement $\Delta \delta_d$. Consequently, regulating the span between the beam's supports we can not only regulate the rigidity of the dynamometer, but model certain values of coefficient of elastic resistance of soil massif, which determines its reaction to dilatancy, i.e. taking into account (2.2), (2.3) and (2.6):
– for cylindrical shear surface

$$S = \frac{EA}{(1+\nu)r}, \qquad (3.2)$$

– for flat shear surface

$$S = \frac{EA}{(1-\nu^2)\omega b}, \qquad (3.3)$$

where A is the area of shear surface in the test.

In the same way it is possible to model the required values of modulus of elasticity of the massif.

During the tests it is possible to use also standard calibrating dynamometers. Conducting tests with three or more standard dynamometers with various stiffness allows to establish functional dependence of resistance to shear on the degree of dilatancy constraint.

Instruments DACS-1 and DACS-2 were used by the author in the course of this research. They ensure step-by-step application of the shearing force, which allows to model conditions of gradual loading of a construction in soil. When it is required to measure resistance with constant shear speed, the force can be applied to carriage 5 by means of an electric motor furnished with a reduction gear and a timing device controlling the movement.

This is the procedure of DACS-apparatus testing. A given normal pressure σ_{n_o} is generated on the surface of sample (10) by screw jack (12) via dynamometer (13) with a fixed rigidity and piston (15). The shearing force is applied step-by-step to carriage (5) with plate (17) attached to it. Soil grains or other material is glued onto the plates. At each step the mobilized value of resistance to shear τ_i (instant and residual) are measured, and the change of force in dynamometer (13) caused by dilatancy or contraction (dilatant stress $\Delta \sigma_d$). In this way resistance of soil to shear is determined not by initial pressure σ_{n_o}, but aggregate ultimate pressure $\sigma_{n_o} + \sigma_d$. In the limiting equilibrium state, static resistance to shear τ_u^{st} is measured first, and when it is overcome, kinematic resistance to shear τ_u^k is measured while the carriage is moving with a constant value of the shearing force. In extra-precise experiments the increment of stresses in the shear plane is registered by pressure and its increment due to dilatancy, shearing force, shearing movement, upward and downward piston travel.

The DACS apparatus also allows to conduct tests according to the standard method, i.e. in the conditions of free dilatancy. To do this, constant normal pressure on the sample is maintained in the process of shear by means of jack (12).

3.3 DILATOMETRIC APPARATUS OF DIRECT SHEAR (DADS)

The design of this instrument corresponds to the phenomenological model described in Section 2.3 (Fig. 2.4). The appearance of the instrument is represented in Fig. 3.4. The peculiarity of the DADS is that normal pressure is applied through regulated dynamometers from both sides of the sample. This ensures the possibility to model the conditions of shear in soil massif, including the shear along the surface dividing volumes with different values of deformational characteristics. To do so, rigidity values are calculated separately for the upper and lower dynamometers, while the general deformativity of the massif is characterized by the given above values of modulus of elasticity E_{red} and coefficient of elastic resistance K_{red} (see Section 2.3).

Figure 3.4. Dilatometric Apparatus of direct shear (DACS).

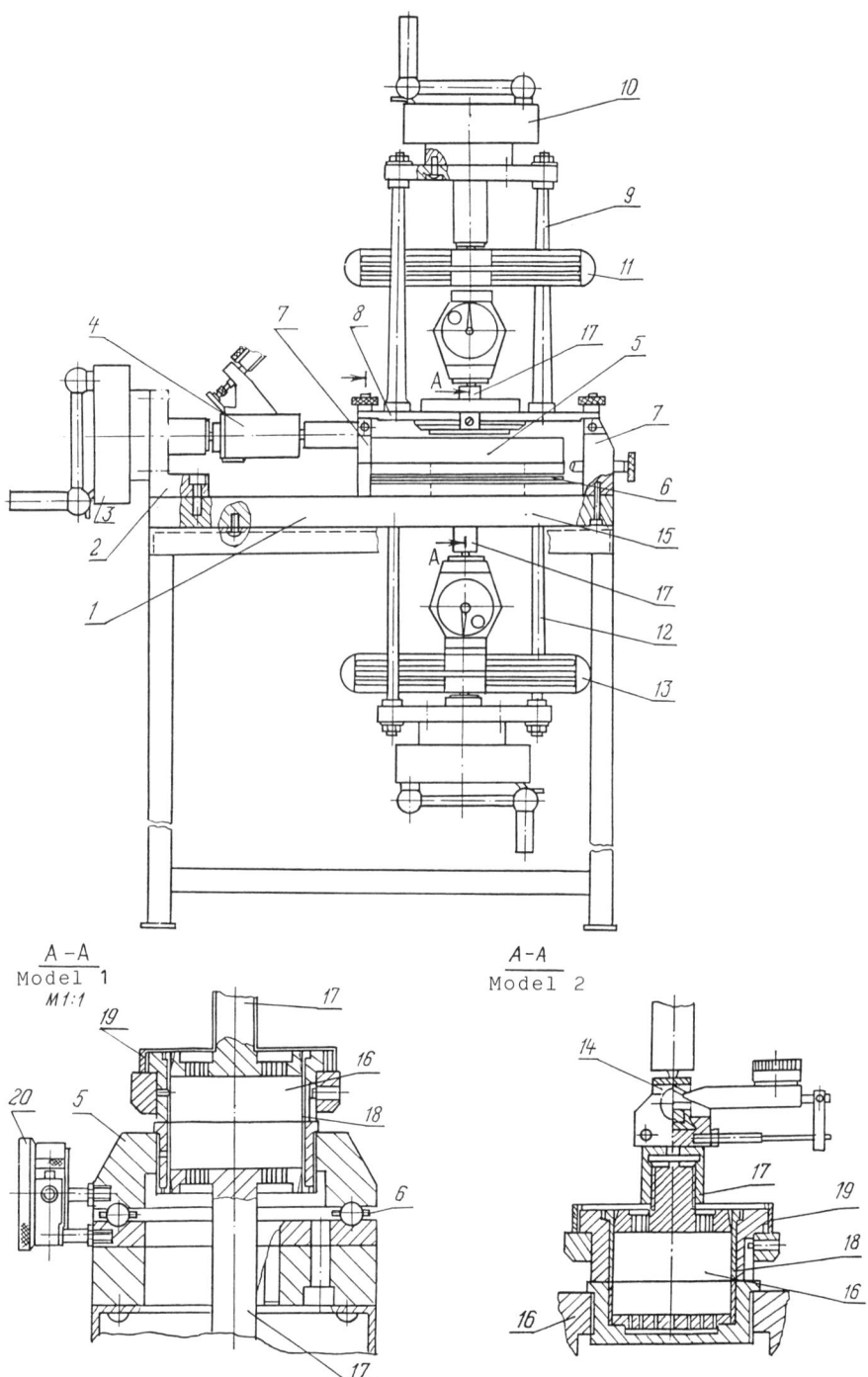

Figure 3.5. DADS design. 1. slab; 2. rest; 3. jack; 4. dynamometer; 5. movable lower carriage; 6. bearings; 7. rests; 8. upper carriage; 9. vertical jack frame; 10-11. dynamometers; 12. frame of the lower jack; 13-14. regulated dynamometers; 15. oval hole; 16. samples; 17. pistons; 18. box; 19. regulating ring; 20. indicator.

The design of the DADS apparatus is represented in Fig. 3.5. The instrument is installed on a slab (1), on which there is a rest (2) for applying the shearing force from a screw jack (3) via a dynamometer (4); movable lower carriage (5) on bearings (6); rests (7) for fixing upper carriage (8) with frame (9) for screw jack (10) and normal-pressure dynamometer (11). Frame (12) for the second normal-pressure screw-jack (13) and regulated dynamometer (14) is also fastened onto the lower carriage. For this purpose oval holes (15) are made in slab (1), which allows frame (12) to move parallel with carriage (5). At both ends of sample (16) there are conical pistons (17) for applying normal pressure (their form decreases friction of the sample against the walls of box (18). Box (18) consists of two halves, the clearance between which is set by king (19). Horizontal movement of the carriage is monitored by a clock-type indicator (20) with a precision of 0.01 mm. In the conducted tests standard ring boxes for samples with an area of 40 cm^2 were used.

This is the test procedure. Given normal pressure σ_{n_o} is generated at the ends of sample (16) from above and from below by screw jacks (10) and (13) through regulated (or calibrating) dynamometers. While modelling different deformativity of the massif lying on both sides of the shear surface, the rigidity of dynamometers is set in accordance with the values of coefficient of elastic resistance of solid K and modulus of elasticity E. The shearing load is applied to carriage (5) step-by-step. At each step the mobilized value of resistance to shear τ_i (instant and residual), the increment of normal pressure $\Delta\sigma_d$ due to dilatancy, and the corresponding movement $\Delta\delta_{d_1}$ and $\Delta\delta_{d_2}$ in the direction of each dynamometer are registered.

In the limiting equilibrium state the static resistance to shear τ_u^{st} is measured, and when it is overcome, the kinematic resistance to shear τ_u^k is measured in the process of shear with constant speed.

3.4 SPECIAL TESTING METHOD WITH THE USE OF THE SERIAL SHEAR APPARATUS BCB-25

As research indicates, shear tests with acceptable precision of the obtained parameters can also be conducted with serial instruments of the BCB-25 type, provided that vertical pressure is applied through a regulated dynamometer (Fig. 3.6). The rigidity has to be set according to the given value of coefficient of elastic resistance of massif K_{red}, calculated by the Equation (2.16).

When it is necessary to establish functional dependence of resistance to shear and strength parameters on deformative characteristics of the massif, it is possible to conduct tests with consecutive installation onto the piston of standard calibrating dynamometers of various rigidity. While doing so, unlike recommended in the instruction to the BCB-25 type instruments, the constant normal pressure on the sample should not be maintained.

42 *Soil testing in the conditions of constrained dilatancy*

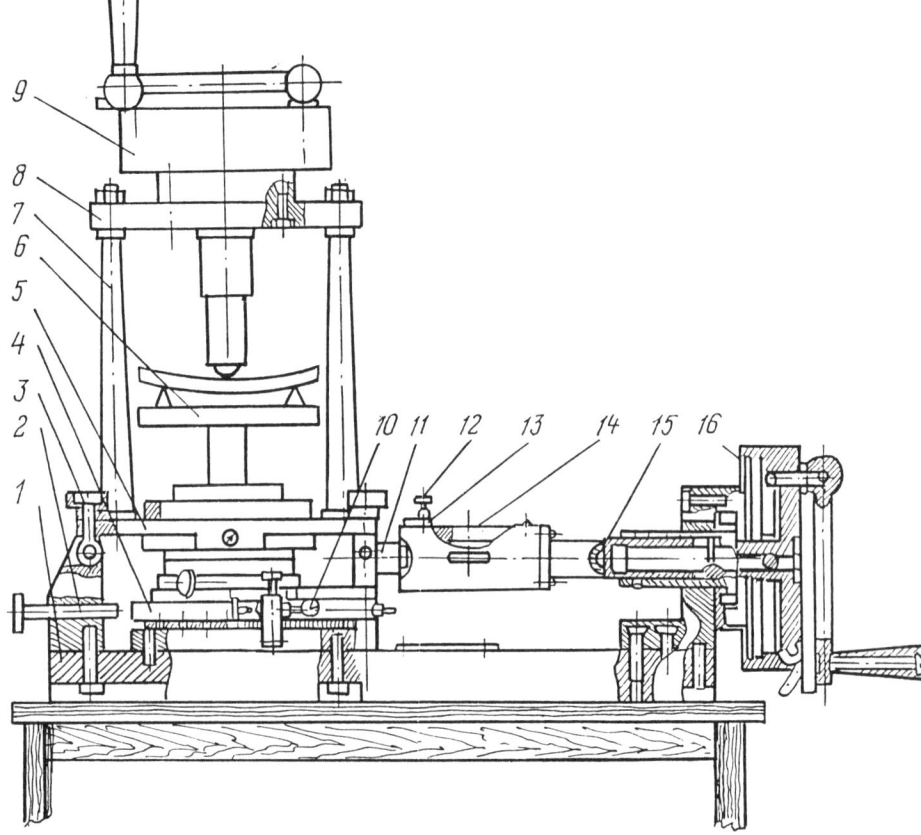

Figure 3.6. Modernized BCB-25 instrument design. 1. slab; 2. regulating screw; 3. installation screw; 4. movable carriage; 5. upper carriage; 6. regulated dynamometer; 7. post; 8. jack installation slab; 9. jack; 10. movement indicator; 11. pusher; 12. fixing screw; 13. jack; 14. indicator; 15. pusher; 16. jack.

3.5 DILATOMETRIC APPARATUS OF CONTACT SHEAR OF REINFORCING ELEMENTS (DACS-A)

This instrument, officially titled 'Device for establishing tribo-technical parameters in the process of friction of a soil sample against an element reinforcing the sample' (Sobolevsky & Popov, 1989), has received a USSR copyright certificate number 1658038. The design of the instrument corresponds to the model represented in Fig. 2.3.

The DACS-A instrument is designed for contact shear tests of materials and geomembranes, reinforcing soil (geotextiles, films, metal bands etc.). This instrument allows to measure the value of contact friction not only as a function of reinforcing element surface character (coarseness), but taking into account its

Dilatometric apparatus of contact shear of reinforcing elements (DACS-A) 43

linear deformativeness (stretching under load). The measured resistance to shear depends on the ratio of deformative characteristics of soil on the one hand, and of material placed therein on the other; in other words, on their interaction. Two failure mechanisms possible in this case have been considered: consequent to the 'material-soil' contact, and as a result of discontinuity or free-flowing nature of the material.

The schematic design of the instrument is represented in Fig. 3.7. DACS-A consists of a slab (1) onto which a rectangular sample box (2) is fastened; and two frames (4) (upper and lower) with screw jack (5), dynamometer of regulated rigidity (6) and pistons (7) for applying normal pressure to both ends of the sample. There are slots in box (2) for placing the tested material (8). The application of shearing force to this material is effected by means of block (9) with steel cable (10) and loading platform (11). The cable is equipped with beam (12) for fastening the tested material (3). The area of contact shear is maintained constant due to the fact that the tested material is 20-30 mm longer than the box.

The horizontal movement and stretching of the reinforcing element is registered by cells (15) and (16), which are installed in front of, and behind the box (2) (at points of entrance and exit of the material into and out of the soil sample). The compaction of the sample at the lower piston is ensured by fixing

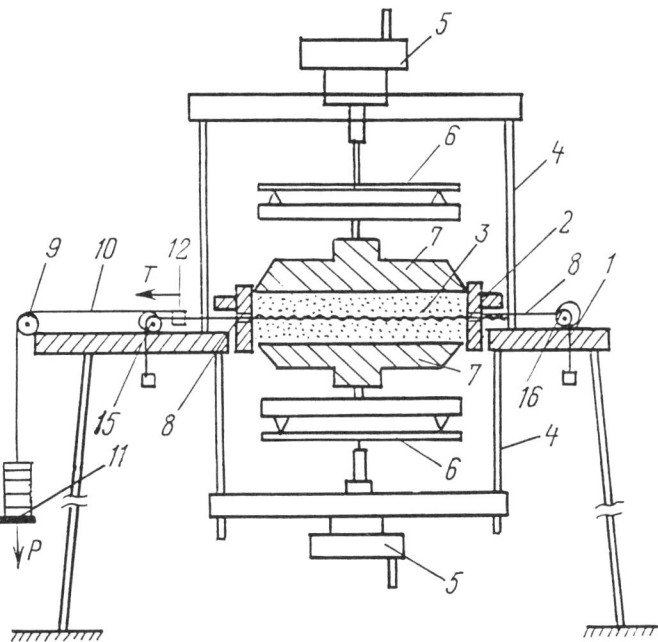

Figure 3.7. DACS-A instrument design. 1. instrument table; 2. sample box; 3. testing material sample; 4. posts of vertical pressure frames; 5. movement cell; 6. regulated dynamometer; 7. piston; 8. cable of cell; 9. loading block; 10. cable; 11. loading platform; 12. beam.

screws which rigidly fasten the lower piston (7) during the preparation for the test.

This is the testing procedure. The lower loading frame, with a regulated dynamometer (6) and a piston rigidly fastened in its initial position, is installed. The box is filled with soil up to the through slot (8), the soil is then compacted to the degree required in the test. The tested material (3) (for example, geotextile, film) is inserted into slot (8). Then the top section of the box is filled up with soil, the soil is compacted, the upper loading system (consisting of a piston and dynamometer) is mounted.

After this the screws fixing the lower piston (7) are loosened, and initial normal pressure σ_{n_o} is generated at the ends of the sample from the top and from the bottom. Cable (10) with beam (12) is installed, and the shearing force is applied step-by-step until failure takes place due to the break-down of contact friction or to the discontinuity or free-flowing of the material.

In the course of the test dilatancy and increment of normal pressure are registered according to the indicators of dynamometers, while the horizontal movement of the reinforcing element is noted by two cells – in front of and behind the sample. The difference of readings allows to establish the reinforcing element's own lengthwise deformation due to the stretching force.

When constrained dilatancy conditions are present in the reinforced soil, respective rigidity is set on regulated dynamometers according to (2.6). When failure takes place within a free-dilatancy pattern (large-area continuous reinforcement, geomembranes), the dynamometers ensure the transfer of only normal pressure which is maintained at a constant level throughout the test (see Chapter 12).

One of the advantages of the instrument is a possibility to shear-test various types of soil. For instance, peat can be placed at the underside of the membrane or cloth, while sand or coarse-grained soil or fill can be placed on top of it. This allows to model the performance in soil of various kinds of material, when drainage layers or antifiltration membranes are introduced.

The instrument gives a possibility to optimize the parameters of materials introduced into the soil by studying the influence of the material's deformative properties (limit of fluidity, stretching deformation, elasticity, resistance to rupture, character of the surface) on the contact friction. This opens up new perspectives for a complex study of reinforced soil as a composite material. Up to now assessments of its behaviour have been made on the basis of the data obtained in the course of separate tests conducted for each material.

3.6 DILATOMETRIC TRIAXIAL APPARATUS (DTA)

The soil failure conditions with complex stress condition, described above, are modelled in the dilatometric triaxial apparatus (DTA). According to Sidorov's

Dilatometric triaxial apparatus (DTA) 45

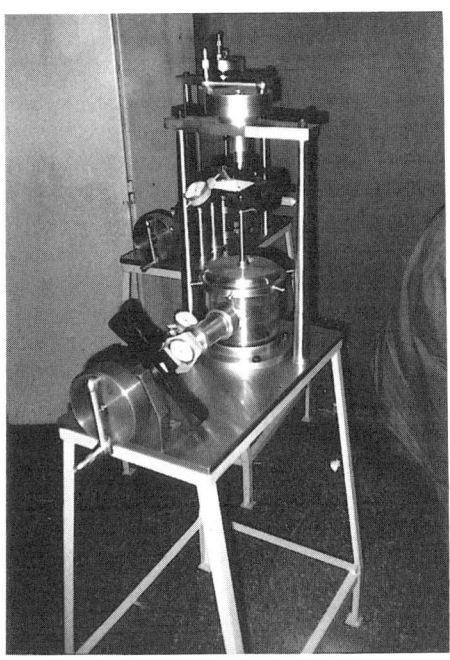

Figure 3.8. Dilatometric Apparatus of Triaxial compression (DTA).

(1972) classification, it is one of the stabilometers of type A. It differs from ordinary instruments in its system of lateral pressure transfer, which in this case is represented by a regulated dynamometer or a set of separate calibrating dynamometers analogous to those described for shear instruments DACS.

The appearance of the DTA instrument is given in Fig. 3.8, and this design is represented in Fig. 3.9b. The instrument consists of camera (1) made of thick organic glass, which is fastened by glue and screwed onto slab (2). The slab is filled with a recess for sample alignment. Camera (1) is topped by a lift-off lid (3) screwed onto the camera. Lid (3) has a hole in the centre filled with gaskets for rod (4) of vertical pressure piston (5). Rod section area is designated to be 1/10 of the sample (6) section area, which is equal to 25 cm^2, which allows to lessen the influence of the rod pressing in on the stress condition.

Lateral pressure on sample (6) encased in latex (7) is generated by the liquid in camera (1). The pressure transfer is effected by lateral pressure piston (8) manufactured in the form of a cylinder. Rings (9) made of polyurethane ensure the system is leak-proof with minimum friction. Plug (10) serves to blow off the air from the camera. For measuring pore pressure there is a hole in slab (2). The hole is fitted with coupling (11) which is joined with manometer (12) or is plugged with a cap.

The loading system consists of instrument table (13) onto which slab (2) is fixed rigidly. Frame (14) for normal pressure screw jack (15), and rest for lateral pressure jack (16) are mounted on table (13). The force applied to pistons (5) and

46 *Soil testing in the conditions of constrained dilatancy*

(8) of the instrument is transferred by regulated dynamometers or calibrating dynamometers with various rigidity. The travel of pistons (5) and (8) is measured by clock-type indicators (17) with a precision of 0.01 mm.

This is the testing procedure. Sample (6) is lowered into camera (1) and mounted in the centre of slab (2). The camera is filled with degassed water or glycerine, after which lid (3) with rod (4) and vertical pressure piston (5) is screwed on. Next the loading system and the movement measuring system are assembled. The rigidity of the lateral dynamometer which models elastic resistance to dilatancy is set according to the calculation made on the basis of expression (3.2). The other dynamometer ensures the measuring of the applied force.

At the beginning of the test the sample undergoes hydrostatic compression $\sigma_{o_1} = \sigma_{o_2} = \sigma_{o_3}$. Then the test is continued while vertical pressure is increasing ($\sigma_1 > \sigma_2 = \sigma_3$) or lateral pressure is decreasing ($\sigma_1 = \sigma_2 > \sigma_3$). In the former case the condition of maximum limit (passive) state, in the latter case the condition of minimum limit (active) state is modelled. The test is continued until

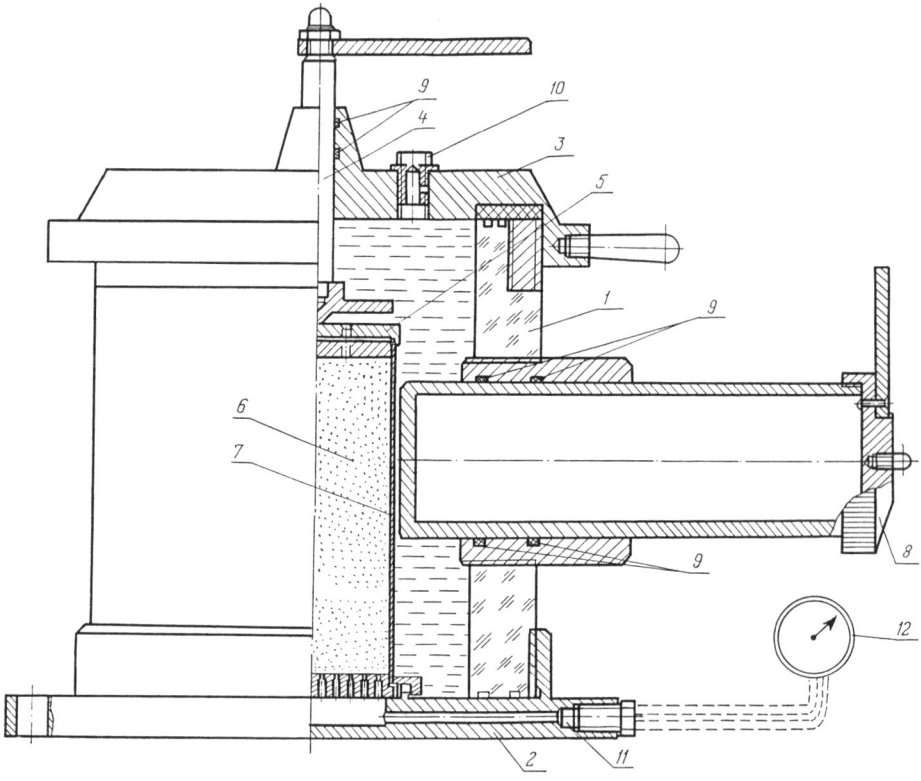

Figure 3.9a DTA instrument design: 1. camera; 2. slab; 3. lid; 4. rod of vertical pressure piston; 5. pressure piston; 6. sample; 7. casing; 8. lateral pressure piston; 9. rings; 10. plug for blowing off the air; 11. coupling; 12. manometer.

Dilatometric triaxial apparatus (DTA) 47

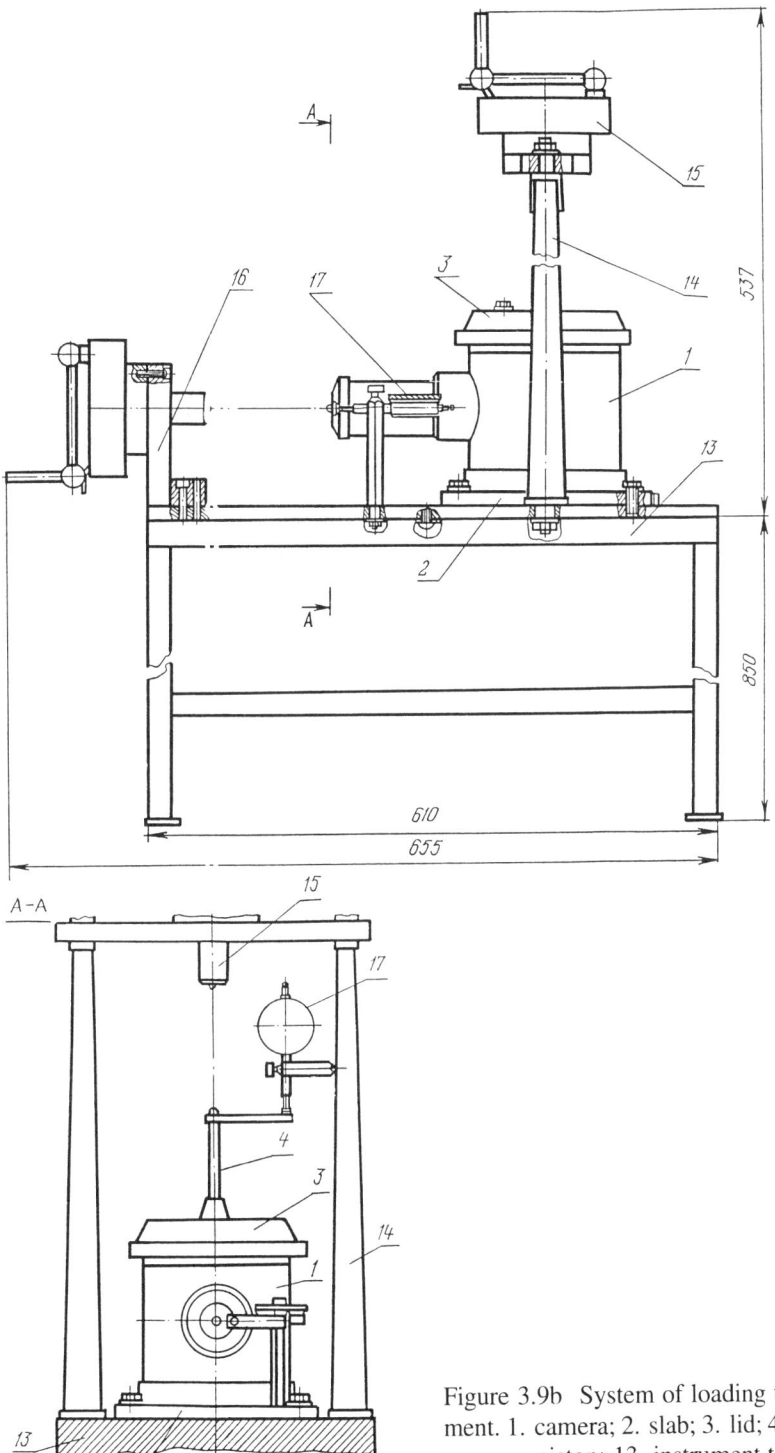

Figure 3.9b System of loading the DTA instrument. 1. camera; 2. slab; 3. lid; 4. rod of vertical pressure piston; 13. instrument table; 14. frame; 15. jack; 16. rest; 17. indicator.

there is thrust or loss of sample stability. A sine qua non for ensuring the purity of constrained dilatancy method tests is total lack of leaks of liquid from the camera, because the slightest leak distorts the pattern of massif's reaction to volume deformations of failure area. This is a complicated task because of high pressures which develop in the camera after the sample is loaded. In the suggested construction the problem was solved by bringing down to minimum the number of joints, and by using effective gasketing. The use of high-viscosity liquids is also preferable.

The DTA instrument can be used to conduct tests by traditional methods, i.e. when there is constant lateral or vertical pressure on the sample all throughout the test. To perform this, one of the dynamometers has to be employed only for measuring the acting pressure, whose value is kept constant by a screw jack. When piston travel takes place in this case, only volume deformations of the sample are registered.

3.7 CONCLUSION

Instruments which go under the generic term 'dilatometric' have been designed and constructed on the basis of elasto-plastic models; according to these models elastic deformations occur in a massif where no mutual grain repacking is observed, while plastic deformations are localised along the surfaces (bands) of sliding or in bulge areas. Constraint of volume deformations, which accompany plastic deformation of grainy medium, is modelled by dynamometers with a rigidity corresponding to modulus of elasticity of massif and to the size of the shear surface (failure surface), i.e. to the characteristics combined in the coefficient of elastic resistance K.

In the absence of constraint on dilatancy the above-described instruments can ensure standard testing methods when initial normal pressures on the sample are constant, which is, in effect, modelling one specific condition for soil failure.

The approach proposed here for designing dilatometric instruments can serve as the basis for creating other special devices for soil testing which would serve to model specific failure conditions.

CHAPTER 4

Contact shear in conditions of constrained dilatancy

4.1 GENERAL PROPOSITIONS

The research data examined below were obtained in the course of a series of tests employing the DACS (Dilatant Apparatus of Contact Shear) and conducted by the constrained shear test method (Section 3.2). The experiments listed herein explore the manner in which shear resistance is influenced by the following major factors:
– degree of dilatancy constraint represented by the elastic resistance coefficient K;
– grain size determined by the mean soil grain diameter d_{50};
– density characterized by the index of density I_D;
– moisture.
According to Terzaghi & Peck (1985)

$$I_D = \frac{e_{max} - e}{e_{max} - e_{min}} \qquad (4.1)$$

where e_{max}, e_{min} are soil porosity coefficients in maximum dense and loose conditions respectively; e is soil porosity coefficient.

The testing of samples was conducted with the three main values of elastic resistance coefficient K equalling 44, 1208, and 2845 MN/m³. These values corresponded to rigidity of the standard calibrating dynamometers with limit loading equalling 2, 10, and 30 kN. Control tests were also conducted with a number of intermediate values of K modelled by a dynamometer with regulating rigidity.

Quartz sands ranging from gravel to silty were used in the research. The bulk of the tests were conducted with artificial (controlled) mixes of various natural sand size fractions characterized by mean grain roughness. The sands employed were coarse, medium, fine and silty. The non-homogeneity curves of these sands are represented in Fig. 4.1. The mean grain diameter d_{50} constituted 1.50 mm for coarse sand, 0.50 mm for medium sand, 0.25 mm for fine sand, 0.10 mm for silty sand.

The physico-mechanical parameters of the sands being studied, calculated by standard methods, are represented in Table 4.1 (GOST, 1975, 1978, 1979, 1982).

The tests were conducted with a triple replication at minimum. The minimal squares method was employed while plotting graphic functions and determining the strength parameters. The experimental values of the control-mix sands strength parameters are represented in Table 4.2.

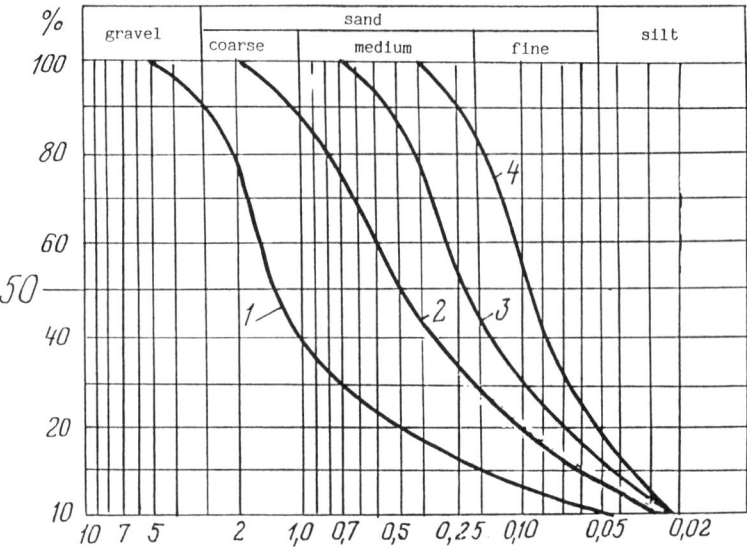

Figure 4.1. Non-homogeneity curves of sands under study. 1. coarse sand; 2. medium sand; 3. fine sand. 4. silty sand.

Table 4.1. Physico-mechanical parameters of control-mix sands.

Sand	Average grain diameter d_{50}, mm	Humidity w_1 %	Density ρ, g/cm³	Coefficient of porosity e	Index of density I_D	Specific cohesion C, kPa	Angle of internal friction under free dilatancy φ_o, grades
Coarse	1.50	5	15.4	0.69	0.3		40.4
			17.4	0.58	0.6	0	41.0
			18.5	0.50	0.8		43.0
			19.5	0.43	1.0		44.1
Medium	0.50	5	16.4	0.70	0.3		34.0
			17.5	0.59	0.6	0	35.3
			18.3	0.51	0.8		36.0
			19.2	0.43	1.0		38.0
Fine	0.25	5	16.4	0.71	0.3	0	34.0
			17.5	0.59	0.6	0	34.3
			18.4	0.51	0.8	1.6	35.0
			19.3	0.42	1.0	3.1	35.0
Silty	0.10	5	19.9	0.40	1.0	6.9	34.0

Table 4.2. Experimental values of dilatant pressure σ_d, kPa dilatant component of strength τ_d, kPa and angles of contact friction φ'_c, grades for quartz dry sands based on test data obtained with the DACS apparatus.

Sand	Values of coefficient of elastic resistance K, MN/m³	Index of density I_D										
		1.0			0.8			0.6			0.3		
		σ_d	τ_d	φ'_c	σ_d	τ_d	φ'_c	σ_d	τ_d	φ'_c	σ_d	τ_d	φ'_c
Coarse	0	0	0	44.1	0	0	43.0	0	0	41.0	0	0	40.4
d_{50} =1.50 mm	44	50	47	42.7	36	33	42.4	23	19	40.2	11	9	38.7
	1208	289	202	34.8	206	140	34.0	159	107	33.9	54	28	27.2
	2845	459	266	29.9	326	189	30.2	219	125	29.5	88	40	24.2
Medium	0	0	0	38.0	0	0	36.0	0	0	35.3	0	0	34.0
d_{50} =0.50 mm	44	58	43	36.4	43	30	34.5	27	18	34.5	10	7	33.8
	1208	190	113	31.3	122	74	31.8	69	42	31.2	15	10	28.8
	2845	263	145	28.9	157	88	29.4	100	55	28.7	54	23	23.2
Fine	0	0	0	35.0	0	0	35.0	0	0	34.3	0	0	34.0
d_{50} =0.25 mm	44	48	32	34.3	33	22	33.4	27	17	33.0	0	1	33.4
	1208	158	92	30.1	120	69	29.6	71	40	29.3	11	6	29.0
	2845	213	115	28.4	145	80	28.3	104	54	27.6	38	17	24.0

4.2 THE INFLUENCE OF DILATANCY CONSTRAINT ON RESISTANCE OF SOIL TO CONTACT SHEAR

We shall now consider the test data of dry maximum-density ($I_D = 1.0$) sands which are the most dilatant. The Fig. 4.2 a, b represent the overlapping charts for coarse medium and fine sands plotted on the basis of the data provided in Tables 4.1 and 4.2. The lines '0' feature standard free dilatancy test results (σ_{n_o} = const), the line '1', '2', '3' represent results with elastic resistance coefficient K equalling 44, 1208, 2845 MN/m³ respectively.

The comparison of the obtained functions reveals a considerable difference between the values of resistance to the grain medium shear in the cases of free and restrained dilatancy. In the former case we receive an ordinary chart for the non-cohesive soil, namely a straight line issuing from the point of origin of coordinates and described by the formula (1.3). In the case of constrained dilatancy the shear straight lines mark a certain initial value of resistance to shear on the ordinate, this value increasing with the growth of elastic resistance coefficient K. Thus, the grain medium appears to acquire the property of 'dilatant cohesion'. At the same time a simultaneous decrease of the angle of contact friction φ_c is observed.

The lines '1', '2', '3' of the charts in Fig. 4.2 are described by the formula

$$\tau_u = \tau_d + \sigma_{n_o} \operatorname{tg} \varphi'_c, \qquad (4.2)$$

where the quantity τ_d we defined as the dilatant component of strength. The

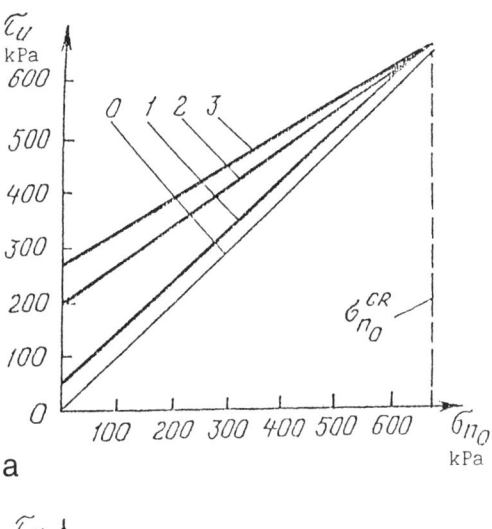

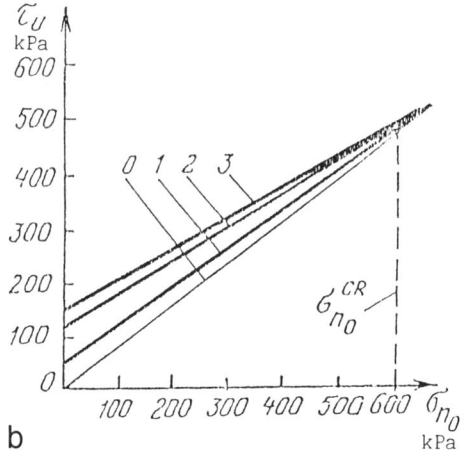

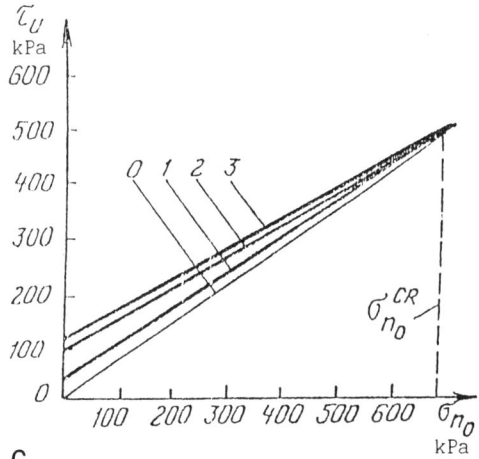

Figure 4.2. The 'ultimate resistance to shear τ_u – initial normal pressure σ_{n_o}, function based on the test data obtained with the DACS appparatus: a) coarse sand; b) medium sand; c) fine sand. 0. under free dilatancy; 1. when $K = 44$ MN/m³; 2. when $K = 1208$ MN/m³; 3. when $K = 2845$ MN/m³

parameters of strength τ_d and φ'_c in the formula (4.2) are functions of the elastic resistance of soil massif coefficient K and, consequently, of the modulus of elasticity E respectively (see Section 2.2).

The appearance of 'dilatant cohesion' is connected with the increase of normal pressure on the sample as a result of thrust when the shear band is widened (see Section 2.1). Consequently, the charts in Fig. 4.2 can also be described by formulas of the equation

$$\tau_u = (\sigma_{n_o} + \sigma_d) \operatorname{tg} \varphi'_c \qquad (4.3)$$

where σ_d is the increment of normal pressure as a result of dilatancy (dilatant pressure).

If we now turn to Table 4.2, we shall observe that value σ_d, similarly to τ_d, is a function of K. Such a description of the obtained functions corresponds to a transfer of axis τ_u to the left by the value of $\sigma_d = \tau_d / \operatorname{tg} \varphi'_c$ (Fig. 4.3). However, in the majority of cases it is more convenient to employ a formula of equation (4.2), treating resistance to shear as a function of initial pressure (σ_{n_o}) and taking into consideration the influence of conditions of dilatancy as τ_d, $\varphi'_c = f(K)$. In such cases the Coulomb law represented in (1.3) can be treated as a specific application of formula (4.2) taking place under free dilatancy condition ($K = 0$, σ_{n_o} = const).

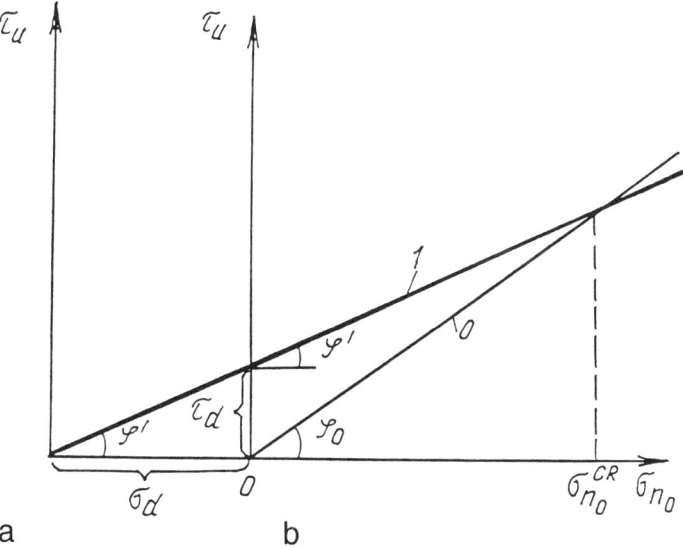

Figure 4.3. Representation of the 'ultimate resistance to shear – normal pressure' functions in the frames of reference: a) $\tau_u = f(\sigma_{n_o} + \sigma_d)$; b) $\tau_u = f(\sigma_{n_o})$. 0. free dilatancy of sand; 1. constrained dilatancy.

54 *Contact shear in conditions of constrained dilatancy*

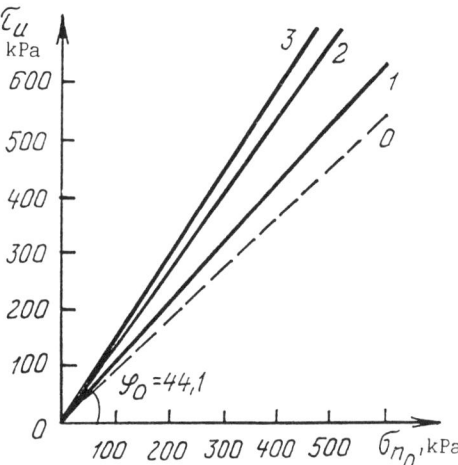

Figure 4.4. Erroneous representation of functions $\tau_u = f(\sigma_{n_o})$ for sands under study. 0. under free dilatancy; 1. when $K = 44$ MN/m³; 2. when $K = 1208$ MN/m³; 3. when $K = 2845$ MN/m³.

If functions $\tau_u = f(\sigma_{n_o})$ are plotted without considering additional normal pressures caused by dilatancy, the value of the angles of internal friction will turn out unexpectedly high (Fig. 4.4). They will constitute (for coarse sand) 46.8° with $K = 44$ MN/m³, 53.8° with $K = 1208$ MN/m³, and 55.7° with $K = 2845$ MN/m³, i.e. exceed 45°. Similar mistakes caused by erroneous interpretation of shear tests, conducted in the conditions of restriction of volume deformation of the sample, can often mislead researchers. For instance, Wernick (1977), employing test data obtained by a shear apparatus with the so-called 'true direct shear', received values equalling 53.1° and more. He supposes the increase in the value of φ above normal to be caused by dilatancy and suggests to allow a correction for the so-called 'angle of dilatancy' while calculating the resistance to shear (see Section 1.7).

Studying the charts represented in Figs 4.2 and 4.3, attention is attracted to a special point corresponding to the intersection of the shear lines '1', '2' and '3' with the line '0'. This point represents the only case of coincidence of shear resistance values in the case of free and constrained dilatancy, which corresponds to the conditions describing the concept of critical density – the constancy of sample volume prior and consequent to failure.

Referring the reader to further chapters for a more precise definition of critical density, we shall note here that the preservation of constancy of soil volume requires a certain initial pressure capable of suppressing dilatancy. We term it as initial critical pressure ($\sigma_{n_o}^{cr}$). Its value can be calculated by the equation

$$\sigma_{n_o}^{cr} = \frac{\tau_d}{\operatorname{tg} \varphi_o - \operatorname{tg} \varphi_c'} \quad (4.4)$$

The influence of dilatancy constraint on resistance of soil to contact shear 55

To the left of $\sigma_{n_o}^{cr}$ the actual resistance to shear will be higher, whereas to the right of $\sigma_{n_o}^{cr}$ it will be lower than expected according to the Coulomb equation (1.3) (see Fig. 4.3). The values of $\sigma_{n_o}^{cr}$ based on test data obtained with the DACS apparatus are represented in Table 4.3 and the chart in Fig. 4.5. The chart demonstrates that the initial normal pressure required for the complete suppression of dilatancy decreases exponentially with the increase of the massif stiffness. The suppression of dilatancy in this case is increasingly supplied by the constrain of the arising thrust.

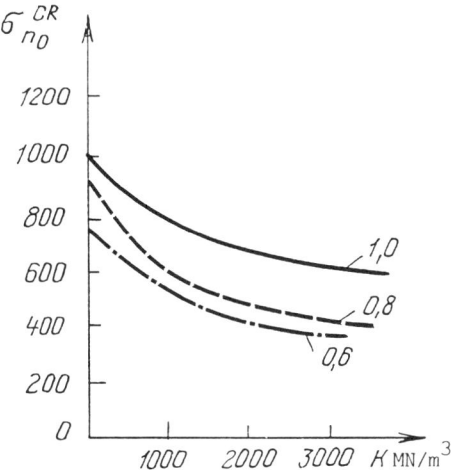

Figure 4.5. The 'critical initial normal pressure $\sigma_{n_o}^{cr}$ – elastic resistance coefficient K' functions with initial values of the index of density for coarse sand $I_D = 1.0; 0.8; 0.6$.

Table 4.3. Values of critical initial normal pressure $\sigma_{n_o}^{cr}$, kPa (complete supression of dilatancy) based on the testing of quartz sands in the DACS apparatus.

Tested sands	Values of elastic resistance coefficient K (MN/m³)	Index of density I_D		
		1.0	0.8	0.6
Coarse	44	1015	903	784
d_{50} = 1.50 mm	1208	737	578	502
	2845	675	539	411
Medium	44	977	764	667
d_{50} = 0.50 mm	1208	652	695	410
	2845	632	540	343
Fine	44	–	539	519
d_{50} = 0.25 mm	1208	663	522	331
	2845	721	495	339

4.3 DILATANT COMPONENT OF STRENGTH AS A FUNCTION OF MASSIF ELASTICITY

The conducted tests revealed that the value of the dilatant component τ_d increased with the increase of the elastic resistance coefficient (modulus of elasticity) of the soil massif adjacent to the shear layer. Fig. 4.6 represents the functions τ_d, $\sigma_d = f(K)$ for dense sands of various grain sizes, plotted on the basis of data from Table 4.2 and from special additional tests wherein intermediary values of the elastic resistance coefficient K were modelled with the help of a dynamometer of regulating rigidity. The values of dilatant thrust stresses σ_d featured in the chart are laid off on the ordinate jointly with τ_d and were calculated as $\tau_d/\mathrm{tg}\,33°$, where the angle $\varphi'_c = 33°$ was conditionally assumed to be the average value for constrained dilatancy.

The maximum increase of τ_d and σ_d occurs at the values of K within the range from 0.0 to 1 000.0 MN/m^3, after which analysis that the experimental curves are adequately enough described by a parabolic function of the equation

$$\tau_d = aK^n \qquad (4.5)$$

The processing of the experimental data base showed that the value of the index n oscillates within the range from 0.29 to 0.36, the average being in the order of 0.33. The coefficient a in the functions represented in Fig. 4.6 reflects the influence of the soil grain size.

Thus the function $\tau_d = f(K)$ can be described by the formula

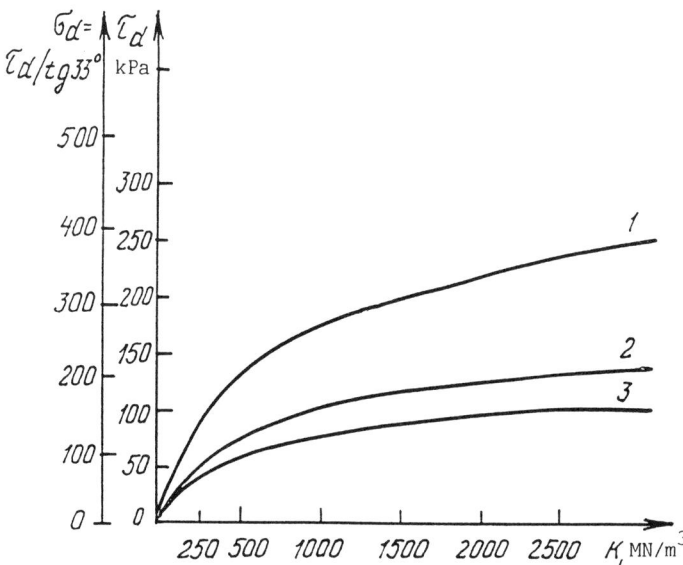

Figure 4.6. Functions τ_d, $\sigma_d = f(K)$ for dense sands. 1. coarse sand; 2. medium sand; 3. fine sand.

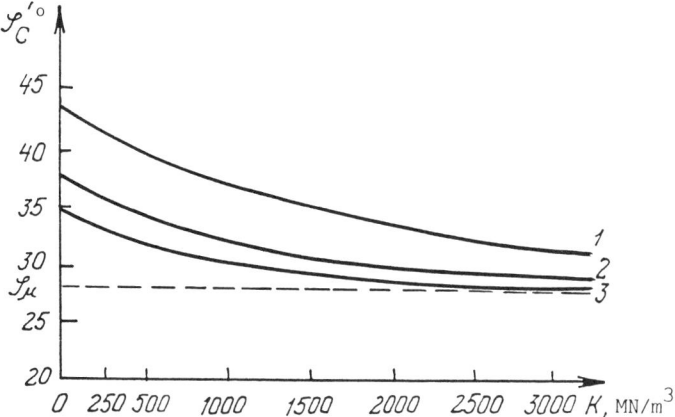

Figure 4.7. Function $\varphi'_c = f(K)$ for dense sands. 1. coarse sand; 2. medium sand; 3. fine sand.

$$\tau_d = a\sqrt[3]{K} \qquad (4.6)$$

or by means of modulus of elasticity taking into consideration (2.2) and (2.3):
– for an axis-symmetric sum (shear of cylindrical surface)

$$\tau_d = a\sqrt[3]{\frac{(1+\nu)r}{E}} \qquad (4.7)$$

– for plane sum (shear of plain surface)

$$\tau_d = a\sqrt[3]{\frac{E}{(1-\nu^2)\omega b}} \qquad (4.8)$$

4.4 ANGLE OF CONTACT FRICTION AS A FUNCTION OF MASSIF ELASTICITY

Constraint of dilatancy for coarse sand, follows from the charts in Fig. 4.3 and Table 4.2, results in the decrease of the value of angle of contact friction, and the more the resistance to dilatancy, the more considerable the decrease of the value of φ'_c.

Let us now consider charts $\varphi'_c = f(K)$, featuring tests conducted with dense sands ($I_D = 1.0$) of control mixes with modelling various values of coefficient K (Fig. 4.7). The plotted curves possess properties of an exponent function. It should be noted that when $K = 0$ the value φ'_c corresponds to that measured with free dilatancy – φ_o.

The constrain of dilatancy leads to asymptotic decrease of φ'_c to a certain value which is constant for the given sand. The recapitulation of the experimental data showed that this value constitutes 24° to 28° depending on the initial density of the sand. It should be remembered that the same value is characteristic for the angle of granular friction of quartz sand (see Section 1.2).

In order to explain the functions $\varphi'_c = f(K)$ we shall apply the Lambe model (Fig. 1.1) and the suggestion to resolve the angle of internal friction into two components: the angle of granular friction φ_μ and the angle of gear φ_g. The former quantity is determined only by the type of the mineral and the physical conditions (moisture, presence of superficially active substances etc.). The latter quantity is a function of physical and mechanical properties of the soil as disintegrated rock material (density, roughness etc.).

So, constraint of dilatancy causes the decrease of the gear component. In other words, the increase in soil stiffness approximates the mechanism of internal friction to the conditions of mineral (granular) friction.

It is possible to object that it is actually the internal gear of grains during the shear that causes their rolling which is manifested in dilatancy. The decrease in the gear component presupposes a decrease in dilatancy. This is actually the way it happens – the maximum dilatancy is manifested in the case of free volume deformation.

The analysis of curves in Fig. 4.7 shows that the maximum decrease of φ'_c can be observed within the range of K from 0.0 to 1 000.0 MN/m³. Thus, the value of φ'_c for coarse sand decreases with $K = 1,000$ MN/m³ by over 8 grades in comparison with the initial value; then the curve begins to slope more gently and the influence of elastic resistance to dilatancy gradually diminishes. With K reaching the values in the order of 3,000 MN/m³, φ'_c practically equals φ_μ.

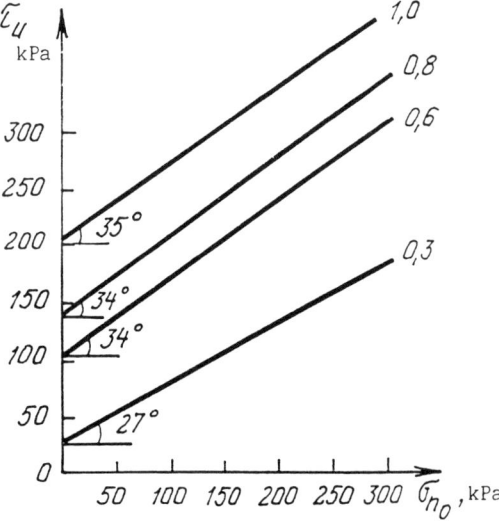

Figure 4.8. Functions $\tau_u = f(\sigma_{n_o})$ for coarse sand when $K = 1208$ MN/m³ and the initial soil density index $I_D = 1.0; 0.8; 0.3$.

The analysis of the experimental data has revealed that the function $\varphi'_c = f(K)$ is best described by an exponential function of the equation

$$\varphi'_c = \varphi_\mu + (\varphi_o - \varphi_\mu)\exp - \alpha K \qquad (4.9)$$

where the value $(\varphi_o - \varphi_\mu)$ is actually the angle of gear designated herein as φ_g; as for the coefficient α, it reflects the influence of grain size (d_{50}) similar to the way coefficient a does in the function (4.6). Taking (2.2) and (2.3):
— for an axis-symmetric sum

$$\varphi'_c = \varphi_\mu + (\varphi_o - \varphi_\mu)\exp - \alpha\frac{E}{(1+v)r}; \qquad (4.10)$$

— for plane sum (shear of plane surface)

$$\varphi'_c = \varphi_\mu + (\varphi_o - \varphi_\mu)\exp - \alpha\frac{E}{(1-v^2)\omega b} \qquad (4.11)$$

4.5 THE INFLUENCE OF INITIAL SOIL DENSITY AND GRAIN SIZE

A decrease in density results in the lowering resistance of the soil to shear. However, in the case of constrained dilatancy this lowering is connected with the decrease not only of the angle of contact friction, but also of the value of the dilatant component of strength τ_d. This is well illustrated by the chart in Fig. 4.8 featuring functions $\tau_u = f(\sigma_{n_o})$ for coarse sand of various density tested with a modelled K equalling 1208 MN/m³.

Analogous charts for functions $\tau_d = f(K)$ and $\varphi'_c = f(K)$ are plotted in Figs 4.9 and 4.10 respectively. The analysis of the first chart shows a considerable decrease of the dilatant component of strength with the decrease of initial sand density. This is symptomatic, because in a looser structure there is a greater possibility of resolving of the dilatant thrust at the expense of re-packing of grains. The initial density exerts less considerable influence on the values of contact friction angles. As is seen from Fig. 4.10, the curves for sands of maximum ($I_D = 1.0$) and mean ($I_D = 0.6; 0.8$) density are plotted close to each other. The only exclusion is the curve for loose sand ($I_D \approx 0.3$). When $K > 1{,}100$ MN/m³ values of φ_c are determined by lower angles of granular friction. Probably it is connected with a possibility for free rolling of grains in the sliding surface in the case of such loose condition.

In a loose soil, contraction results in a partial loss of contacts between grains. Consequently, gear almost disappears and intergranular friction lowers. Thus, with $K = 2845$ MN/m³ the angle of contact friction equalled 24.2° for coarse sand, 23.2° for medium sand and 24° for fine sand, which is several grades less than φ_μ which was conditionally assumed to be 28°.

The analysis of the test data revealed that functions $\tau_d = f(K)$ and $\varphi'_c = f(K)$ for

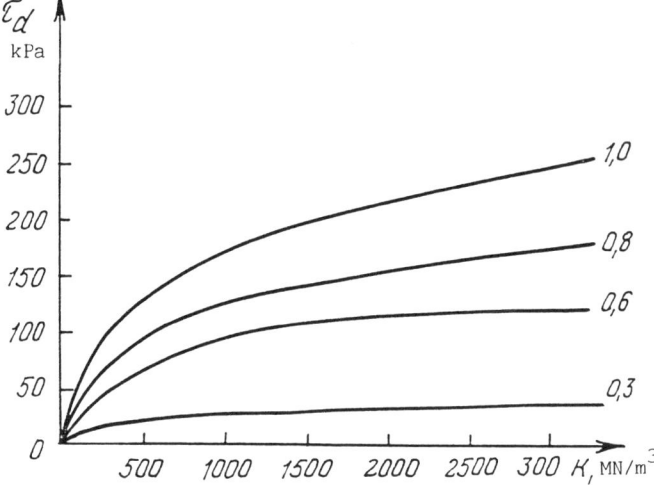

Figure 4.9. Functions $\tau_d = f(K)$ for coarse sand when the initial soil density index $I_D = 1.0$; 0.8; 0.6; 0.3.

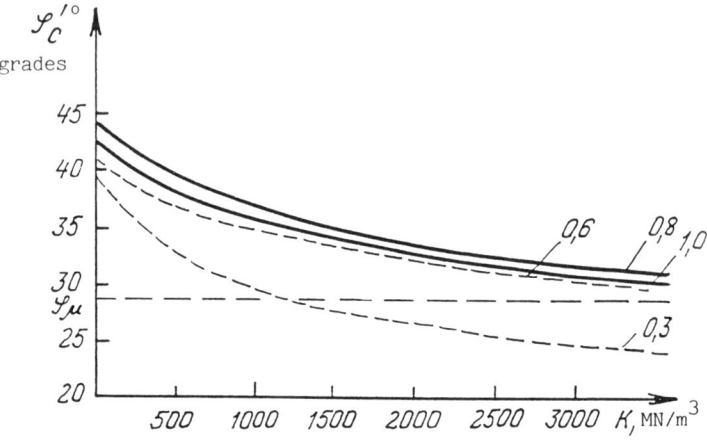

Figure 4.10. Functions $\varphi'_c = f(K)$ for coarse sand when the initial soil density index $I_D = 1.0$; 0.8; 0.6; 0.3.

all control sands within the range of values of $I_D = 0.6$-1.0 are well described by formulas (4.6) and (4.9); coefficients a and α reflecting now the influence not only of grain size, but also of initial soil density. The values of empirical coefficients a and α based on contact shear tests are represented in Table 5.4.

Let us turn to the analysis of functions $a, \alpha = f(I_D)$ on the basis of the charts represented in Figs 4.11 and 4.12. As it follows from Fig. 4.11, the control mix sands give a practically linear function of coefficient a versus density. This means that the increase of density leads to the increase of the possibility for dilatant thrust realization.

The straight lines in Fig. 4.11 cross the abscissa at points with $I_D \approx 0.21$ for coarse sand and $I_D \approx 0.26$ for medium and fine sands. These values correspond to critical initial density. We must admit, though, that testing loose sands presented certain difficulties. The application of even insignificant normal pressure caused compaction which resulted in the scatter of the received values of strength and in distortion of function $\tau_u = f(\sigma_{n_o})$. Besides, it was rather complicated to ensure a uniform density along the height of the sample. This is the reason why the values of τ_d and φ'_c when $I_D \approx 0.3$, represented in Table 4.2, corresponding rather to average statistic values obtained by processing test data within the range of initial normal pressures from 0.0 to 200.0 kPa. As for experiments with $\sigma_{n_o} = 0$, they revealed either zero dilatancy, or a certain contraction, i.e. with $I_D \approx 0.3$ the soil was in a state close to initial critical density.

The chart $\alpha = f(I_D)$ in Fig. 4.12 also reveals an almost linear function within the range of density index $I_D = 0.6$-1.0. However, it proved impossible to find certain values of α for $I_D \approx 0.3$, because while modelling the values of elastic resistance coefficient exceeding 1,100-1,300 MN/m³, the received angles φ'_c turned out to be less than the granular friction angle and did not fit into the function (4.9).

The functions $a, \alpha = f(d_{50})$ in Figs 4.13 and 4.14 provide additional illustration of dependence of coefficients a and α on density and grain size. Of some interest here is the chart $\alpha = f(d_{50})$ corresponding to the data for dense sand and reflecting a quasi-linear function of coefficient α versus grain size. It will be noted, though, that no special physical sense should be adduced to the empirical coefficients a and α; as for the values represented in Table 5.4, they may need adjustment for sands with a different grain roughness and still more so for a constituting mineral.

The increase of grain size leads to enhancing contact pressures on grain edges. This increases the risk of destruction of grains in contact points in the process of

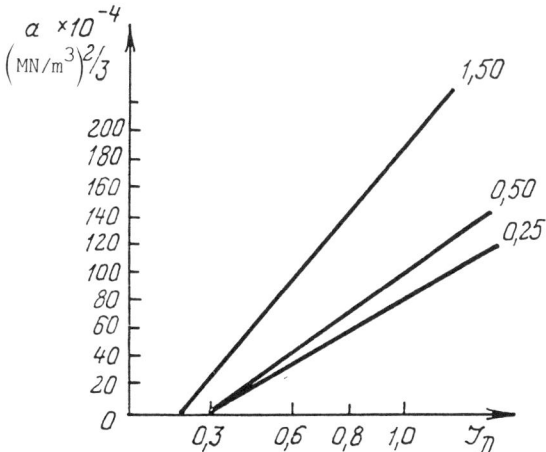

Figure 4.11. Functions $a = (I_D)$ for sands with mean grain diameter $d_{50} = 1.50; 0.50; 0.25$.

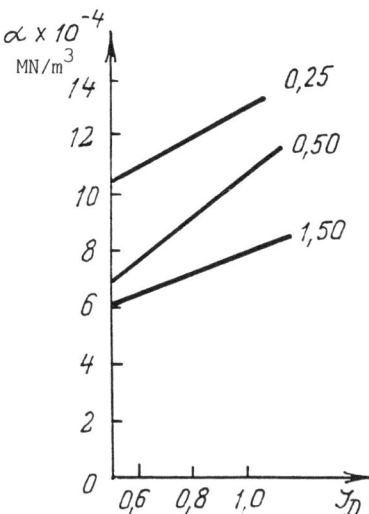

Figure 4.12. Functions $\alpha = f(I_D)$ for sands with mean grain diameter $d_{50} = 1.50; 0.50; 0.25$.

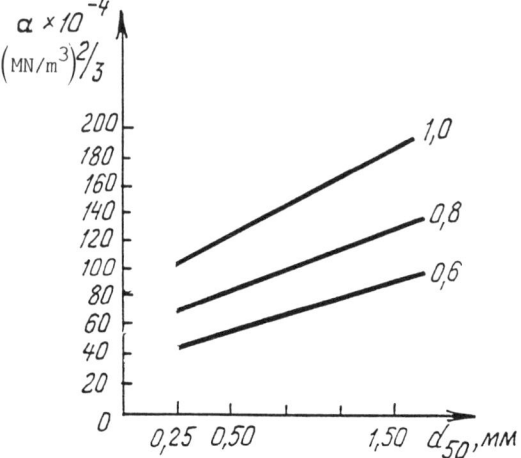

Figure 4.13. Functions $a = f(d_{50})$ for control-mix sands when the density index $I_D = 1.0; 0.8; 0.6$.

developing a dilatant thrust. Such destruction will inevitable influence the obtained values of parameters τ_d and φ'_c.

The lowering of d_{50} to less than 0.15-0.10 mm and the presence in the soil of fractions of silt and clay signifies a change in the character of interaction in the soil, which is especially prominent in the case of increased moisture. As a result of increase in the surface activity, the mechanical gear of grains is substituted by the interaction mediated by hydrate coating. The term 'grain' should then be substituted by the concept 'particle'. That is why the empirical equations (4.6) and (4.9) should be used with care when $d_{50} < 0.15$ mm.

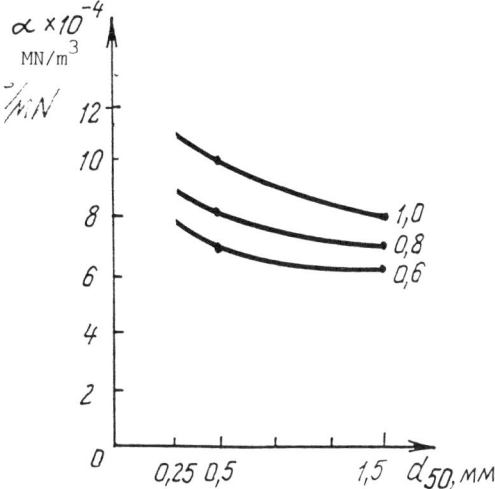

Figure 4.14. Functions $\alpha = f(d_{50})$ for control-mix sands when the density index $I_D = 1.0; 0.8; 0.6$.

4.6 INFLUENCE OF MOISTURE CONTENT

This factor was studied by comparative drained shear tests on sands of various composition in the following states: dry ($w = 3\text{-}5\%$); humid ($w = 10\text{-}12\%$) and saturated. In these tests the carriage of the DACS apparatus was supplied with a plastic tub.

The tests did not reveal the influence of moisture on the values of τ_d and φ'_c for coarse and medium sands, but a decrease of τ_d approximately by 20% and of φ'_c by 16% was registered for fine sand after its full saturation (Fig. 4.15).

A considerably greater influence on strength is exercised by moisture in the case of silty sand with $d_{50} = 0.10$ mm. As is seen from the chart $\tau_d = f(\sigma_{n_o})$ in Fig. 4.16, the increase of moisture to 10-12% leads to a considerable decrease of values of τ_d and φ'_c, while complete water-saturation causes a drop of resistance to shear by 3.0-4.5 times. A comparison of values of σ_d, τ_d and φ'_c for fine and silty dense sand in air-dry, humid and saturated conditions based on test data with K modelled at 1208 MN/m³ is represented in Table 4.4.

As the test results show, a constraint of dilatancy is capable of developing high values of dilatant thrust in the case of dense dry silty sand as well. In this type of soil which possesses a certain cohesion, the failure in the shear band took place not as rolling of separate grains, but as resolving into aggregates of glued particles. It was especially conspicuous when $\sigma_{n_o} = 0\text{-}50$ kPa. We were able to observe soil adhering to the shearing contact carriage plane of the apparatus. As a result, the surface of the shear consequent to the extraction of the destroyed sample was uneven (knobbly). The mechanism of this phenomenon is illustrated by Fig. 4.17.

64 *Contact shear in conditions of constrained dilatancy*

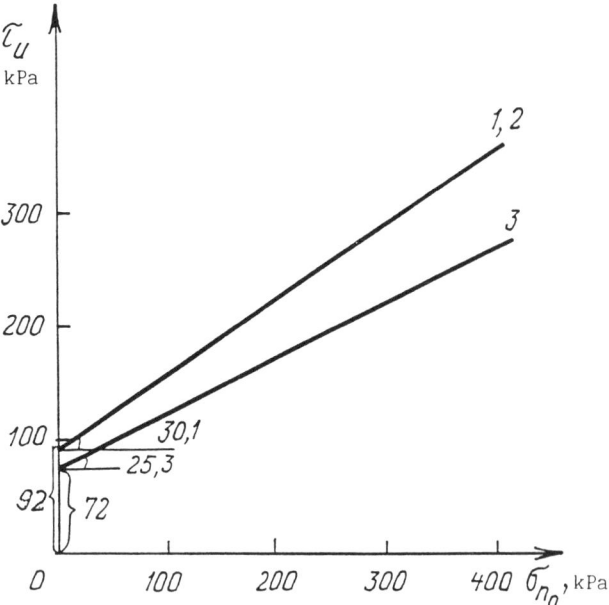

Figure 4.15. Functions $\tau_u = f(\sigma_{n_o})$ for fine sand when $K = 1208$ MN/m³. 1. dry sand; 2. humid sand; 3. saturated sand.

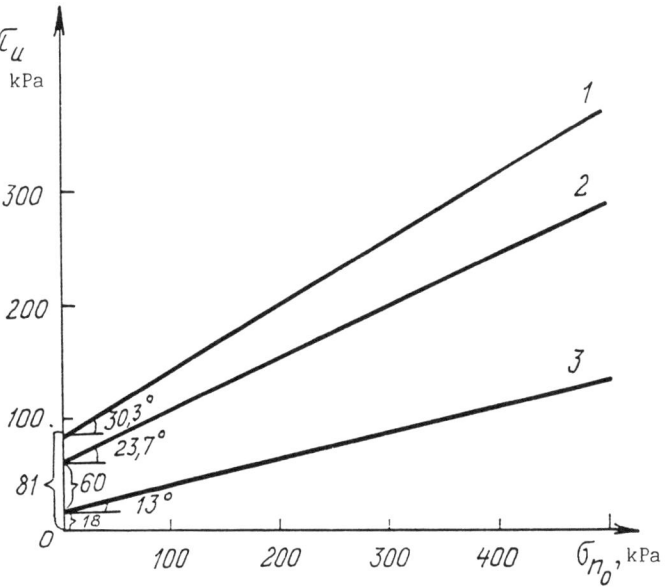

Figure 4.16. Functions $\tau_u = f(\sigma_{n_o})$ for silty sand when $K = 1208$ MN/m³. 1. air-dry sand 2. humid sand; 3. saturated sand.

Table 4.4. Experimental values of σ_d, τ_d, kPa and φ'_c, grades for fine and silty sands with various degrees of humidity ($K = 1208$ MN/m³).

Sand	Humidity w								
	3-5%			10-12%			saturated		
	σ_d	τ_d	φ'_c	σ_d	τ_d	φ'_c	σ_d	τ_d	φ'_c
Fine $d_{50} = 0.25$ mm	159	92	30.1	153	88	30	152	72	25.3
Silty $d_{50} = 0.10$ mm	139	81	30.3	137	60	23.7	78	18	13

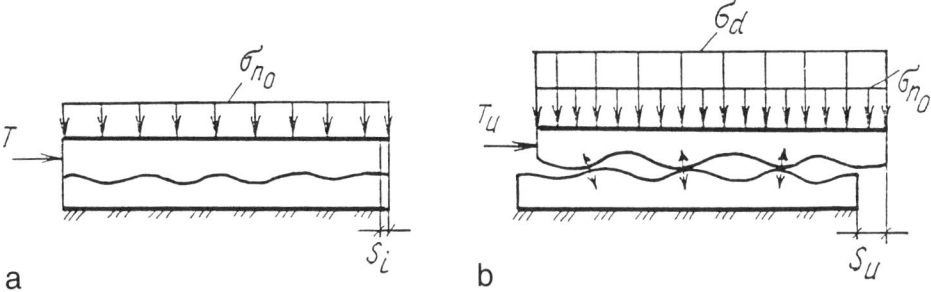

Figure 4.17. Possible mechanism of dilatancy of fine-grained soil possessing the property of cohesion: a) formation of uneven shear surface; b) shifting of knobs with the development of dilatancy.

According to such interpretation, the moistening of silty sand caused the destruction of lumps as a result of loss of cohesion between particles. This process was accompanied by formation of a natural lubrication which preluded the development of gear between sand grains forming soil skeleton.

The tests dealing with silty sands of lower density revealed that its dilatancy is much lower, whereas the scatter of measured values of strength is greater. With $I_D = 0.6$ and lower the soil became extremely sensitive to moistening. Even a slight increase in humidity (up to $w = 10\text{-}12\%$) practically totally excluded dilatancy.

4.7 VALUES OF DILATANT STRAINS

The appearance in the process of shear of dilatant thrust or the dropping of normal pressure is connected with the widening or narrowing of the failure band. Hence it is of interest to analyze the absolute values of dilatancy for the evaluation of the phenomena mentioned above.

Table A1.2 features ultimate values of dilatant strains measured in the course of tests conducted with the DACS apparatus (δ_d^{exp}) and calculated on the basis of data supplied in the Tables 4.2 and 4.4 as

$$\delta_d^{cal} = \frac{\sigma_d}{K} \qquad (4.12)$$

A comparison of δ_d^{exp} and δ_d^{cal} shows that experimental values of strains are lower than calculated values, with the divergency increasing while the rigidity of dynamometer is diminishing. This is accounted for by the loss of a part of dilatant strain due to the friction of the sample against the walls of the apparatus box. Special tests revealed that when the thickness of the sample exceeded 25 mm the measured values of δ_d^{exp} diminished even in the presence of antifriction lubrication.

As the experimental data testify, the maximum values of dilatant strains constitute, as a rule, fractions of a millimetre. this can serve as additional proof that the assumption, regarding the elastic reaction of the massif to the dilatancy of the shearing band (see Section 2.5), is fully justified. Obviously, a compression in the order of fractions of a millimetre, even if it is exerted on loose soil, is not capable of causing any considerable plastic deformations.

Figs 4.18-4.20 represent charts illustrating the dependence of dilatant strain on density, initial normal pressure and elastic resistance coefficient, respectively.

The chart $\delta_d^{cal} = f(I_D)$ in Fig. 4.18 testifies that the diminishing of the density of sand causes an almost linear decrease of dilatant strain and, consequently, of thrust. The same result is received by the increase of normal pressure on the sample (Fig. 4.19). The dilatant trust emerging in the process of the shear also

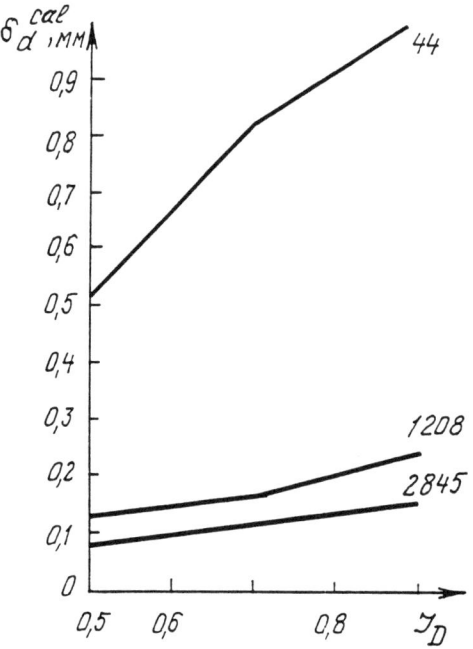

Figure 4.18. Functions $\delta_d^{cal} = f(I_D)$ for coarse sand with modelled $K = 44; 1208; 2845$ MN/l^3.

Values of dilatant strains 67

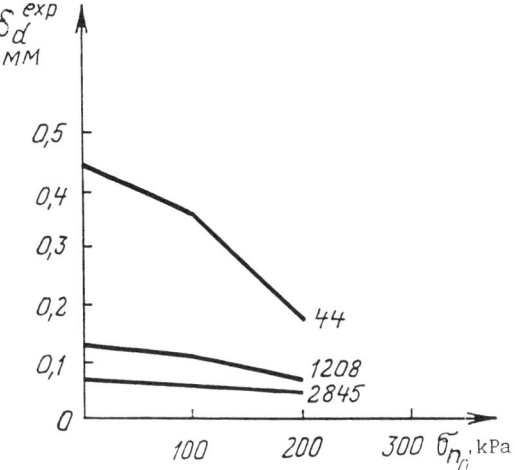

Figure 4.19. Functions $\delta_d^{exp} = f(\sigma_{n_o})$ for coarse sand with modelled $K = 44$; 1208; 2845 MN/m³.

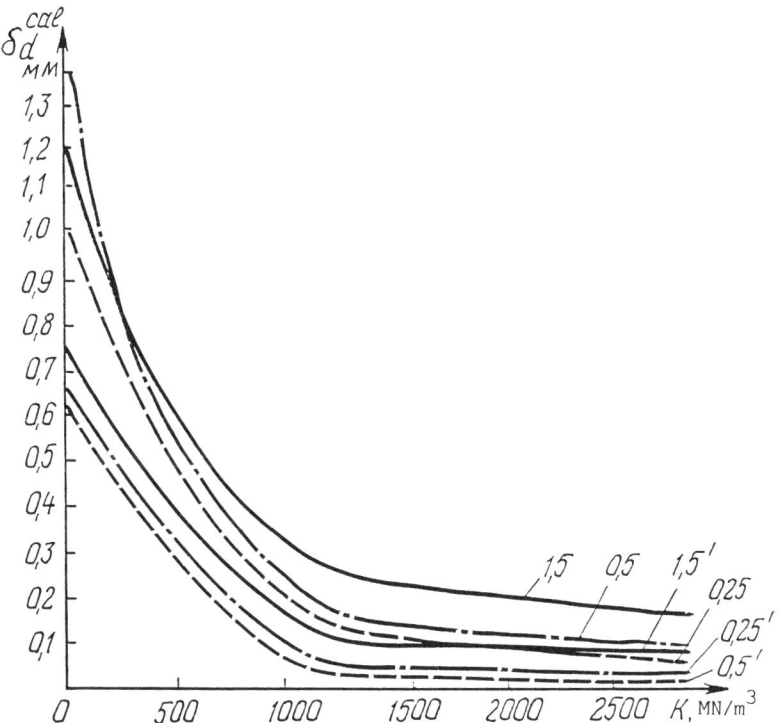

Figure 4.20. Functions $\delta_d^{cal} = f(K)$ for sands with average grain diameter $d_{50} = 1.5$; 0.5; 0.25 mm when the initial soil density index $I_D = 1.0$; 0.6 (marked with dashes).

limits strains proportionally to massif stiffness (Fig. 4.20). The charts $\delta_d^{cal} = f(K)$ shows that the values of strains amounting to little more than 1 mm for coarse and medium sands in the conditions of free dilatancy rapidly drop to fractions of a millimetre with the increase of K.

For total suppression of dilatancy in the failure band there must be generated a pressure which we term as critical. This pressure may either correspond to the initial $\sigma_{n_o}^{cr}$, or develop in the process of dilatant thrust and be ensured by the aggregate of $\sigma_{n_o} + \sigma_d$.

Fig. 4.21 represent chart $\delta_d^{cal} = f(\sigma_{n_o})$ plotted by two points connecting the values of δ_d^{cal} with $\sigma_{n_o} = 0$ and with $\sigma_{n_o} = \sigma_{n_o}^{cr}$. This chart allows to assess the absolute values of dilatancy for sand with various grain size. Furthermore it will be important for the analysis of the influence of the technology for deep foundation construction on the obtained contact friction (Chapters 8-11). So dilatant strains are not great and amount to tenth and hundredth fractions of a millimetre. A greater suppression of dilatancy corresponds to a higher value of resistance to shear and a lower value of volume deformation.

In the analysis of the phenomena of dilatancy and contraction one of the most interesting problems is the thickness of the layer which develops during failure. Table A1.2 features the ration δ_d^{cal}/d_{50}. This ratio does not allow to deduce any empirical function, but it gives a possibility to make an indirect conclusion regarding the influence of the massif stiffness on the thickness of the shear layer.

The considerable decrease of the ratio δ_d^{cal}/d_{50} with the increase of stiffness testifies to an accompanying growth of resistance to the re-grouping of the initial grain structure. As a result the thickness of the contact grain layer diminishes and

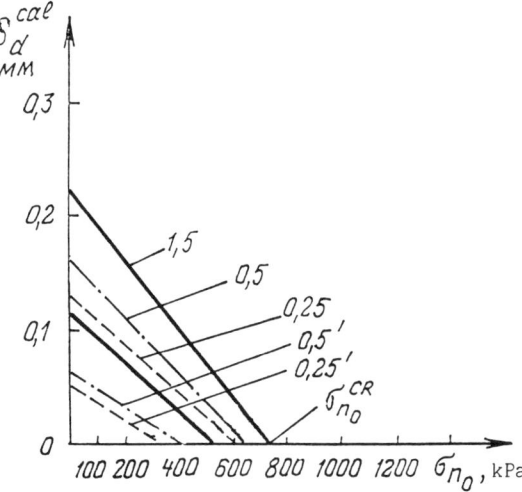

Figure 4.21. Functions $\delta_d^{cal} = f(\sigma_{n_o})$ for sands with average grain diameter $d_{50} = 1.5$; 0.5; 0.25 mm when the initial soil density index $I_D = 1.0$; 0.6 (marked with dashes).

the mechanism of shear increasingly approximates the character of mutual shift of elastic blocks (see Section 1.4).

This last observation certainly bears on dense soils (I_D = 1.0-0.8), because the increase of porosity opens up greater possibilities for re-packing of grains.

4.8 INFLUENCE OF GRAIN STRENGTH

Plastic deformations of non-cohesive soil are also possible as a result of crushing of grains into finer grains due to overstress in contact points. In the case of quartz sands it requires a pressure in the order of 700 kPa. In the presence of less strong minerals, for instance fieldspar, the pressure can be much lower. It should also be taken into consideration that the larger the grains are and the more homogenous is their structure, the larger the concentration of stresses taking place on their edges. Therefore, a due allowance for the influences of failure of grains is especially essential in the case of gravelly and coarse fragmented soils.

It is assumed that pressures on the soil approximating 700 kPa are rarely generated and these cases are limited to soil dams or pile foundation bases. But if generating additional thrust stresses in the condition of constrained dilatancy is taken into account, it transpires that the critical pressure of 700 kPa is easily generated with much lower values of initial normal pressures.

To clarify this point we shall return to chart $\sigma_d = f(K)$ in Fig. 4.6. A dilatant thrust stress σ_d = 370 kPa is created during the shear in coarse dense sand when K = 2,000 MN/m³. According to (2.2) such value of K corresponds to the radius of contact surface r = 0.12 m (e.g. that of a pile) when the modulus of elasticity E = 300 MPa. In this case an initial stress σ_{n_o} = 330 kPa is necessary to generate the pressure of grain failure. Obviously, with grain size exceeding 1.5 mm the ratio between σ_{n_o} and σ_d will prove much less.

During shear testing of sands when the samples were subjected to pressures over 300 kPa there was an audible characteristic crackling accompanying the failure of grains in contact points. When the pressure was raised the crackling became louder, which testifies to the massive character of such crushing. After silting the coarse sand consequent to a large number of tests it transpired that the share of fine fractions grew at the expense of coarser fractions. The massive crushing of grains during shear deformations of coarse fractured soils in the condition of constrained dilatancy must inevitably cause a change in the values of τ_d and φ_c' depending on initial normal pressure. A failure to allow for such phenomena may lead to errors in calculation of functions (4.6) and (4.9).

4.9 CRITICAL DENSITY AND CRITICAL NORMAL PRESSURE

Fig. 4.22 a, b represent several experimental functions characterising the dilatan-

70 Contact shear in conditions of constrained dilatancy

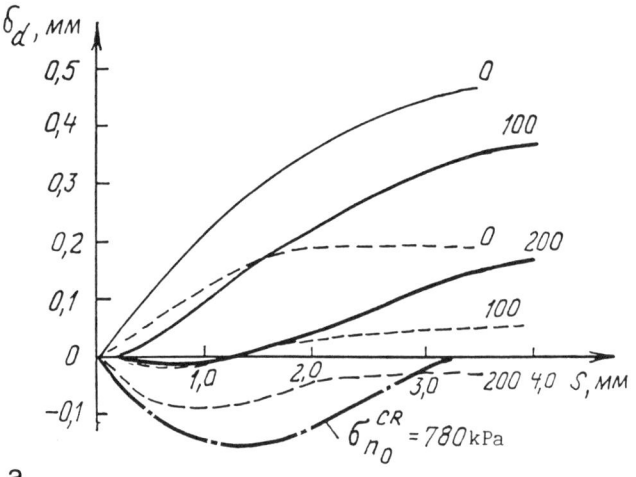

a

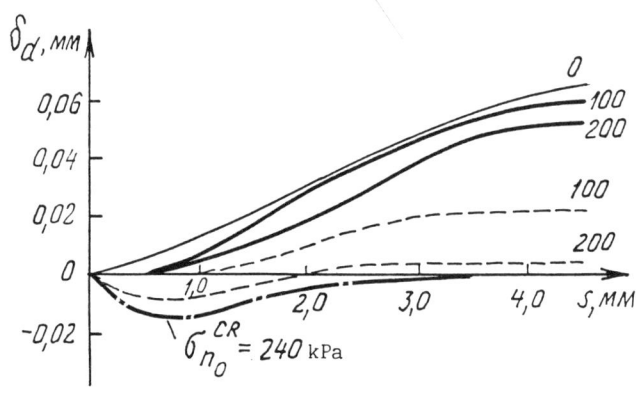

b

c

Figure 4.22. Functions $\delta_d = f(S_i)$ for coarse sand with $\sigma_{n_o} = 0, 100, 200$ kPa when: a) $K = 44$ MN/m³; b) $K = 1208$ MN/m³; c) $K = 2845$ MN/m³ ——— $I_D = 1.0$, ---- $I_D = 0.6$.

cy of coarse and medium sands. Each of the charts testifies to the fact that with the growth of initial normal pressure and the decrease of density, the dilatancy diminishes.

With a certain normal pressure the curves acquire initial loops of contraction which are especially prominent in the case of the sand of average density. This is probably caused by the mechanism of grain gear. In the process of formation of the slip surface grains recompact going into mutual gear. A phenomenon similar to the backlash in pinion teeth is observed.

With a certain initial pressure $\sigma_{n_o}^{cr}$ for each value of K there comes a condition where the soil preserves the same volume prior and consequent to the shear. Thus the condition of critical density is met. The shear curves for $\sigma_{n_o}^{cr}$ are shown in the charts with dashes. When $\sigma_{n_o} > \sigma_{n_o}^{cr}$ dilatancy is substituted by contraction.

The charts in Fig. 4.22 demonstrate clearly that critical density should be necessarily correlated with the stress condition of the soil. Even the most dense soil can be made to contract by certain pressures. But such contraction may then be connected not so much with recompaction, as with grain crushing. As it is seen from Table 4.3, the values of $\sigma_{n_o}^{cr}$ for initially dense sand are close to or exceed pressures of quartz grains crushing defined above as equalling about 700 kPa.

Tests in a contact shear apparatus revealed that ultimate resistance to shear is mobilised after strain ranging from 4.5 to 5.5 mm. In the case of silty and fine saturated sands these strains amounted to the order of 3.5-4.0 mm.

4.10 PEAK AND RESIDUAL STRENGTH OF SANDS

The experiments examined above were conducted by a method where the load was increased step-by-step with pauses for the dying down of shear strain S_i. The functions of mobilised resistance to shear versus S_i in all cases corresponded to curve '2' of the chart in Fig. 4.23.

In the process of shear the ultimate ('static') value of τ_u^{st} was achieved after a

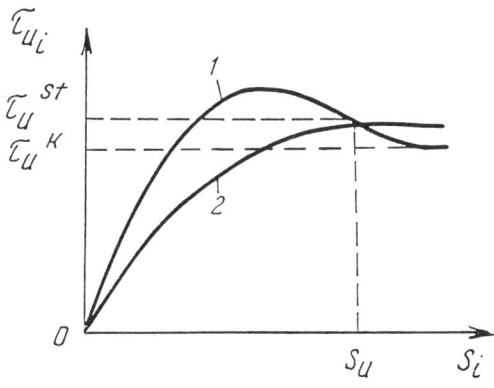

Figure 4.23. 'Static' and 'kinematic' shear curves. 1. abrupt shear; 2. step-by-step loading.

certain strain S_u. If tests were continued with dense sands the value was preserved, as a rule, unchanged, which also testified to preservation of a constant dilatant thrust in the layer of shear. A certain decrease of τ_u^{st} (by 10-15%) was observed in the case of sands with average density ($I_D \leq 0.6$).

Tests with a fixed rate of shear revealed a different character of function $\tau_i = f(S_i)$ only in the case of rapid loading. When the shear was effected by an abrupt push of the carriage this function corresponded to curve '1' in Fig. 4.23, i.e. a certain instant peak value of τ_u could be observed. In the case of further shear it decreased to values which were usually 5-15% lower than τ_u^{st}.

When the rate of shear did not exceed 3 mm/min the character of function $\tau_i = f(S_i)$ did not differ greatly from the case of step-by-step loading. The ultimate 'kinematic' value of τ_u^k was achieved practically by the same strains as the 'static' value.

A possible explanation for the emergence of peak strength in sands could be as follows. In the case of a rapid application of shear load, the response to it is effected by a larger number of grains whose mutual gear prevents them from immediately forming a clearly oriented shear surface. The shear stress is distributed over a thicker layer of grains. As a result of this, resistance to shear registered at the initial moment is higher by 15-20%.

4.11 ON REASONS FOR CURVATURE OF FUNCTION $\tau_u = f(\sigma_{n_o})$

We presume that there are two basic reasons: physical and methodical. The former is connected with the fact that, when dilatancy is constrained, comparatively low values of initial normal pressure (in the order of 300-400 kPa) may result in a failure of grains in the shear band. Correspondingly, the dilatant thrust diminishes and the angle of contact friction decreases.

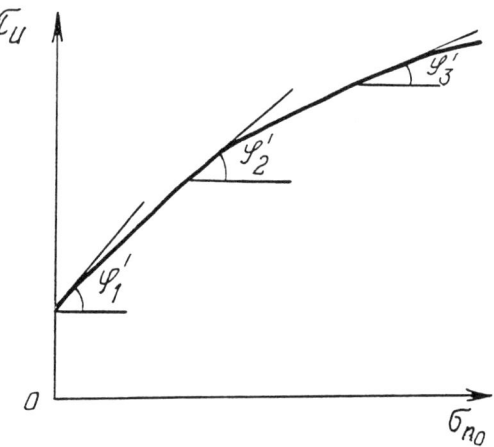

Figure 4.24. Example of representation of function $\tau_u = f(\sigma_{n_o})$.

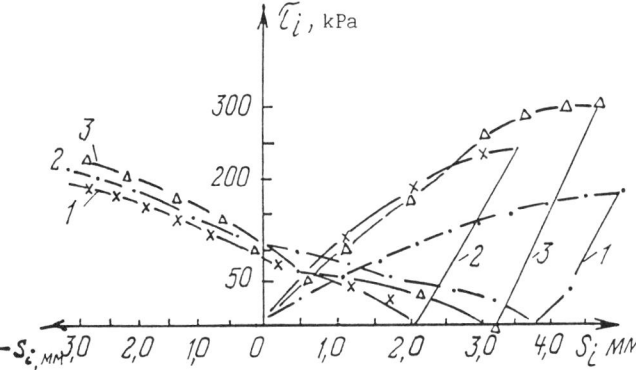

Figure 4.25. Shear curves $\tau_u = f(S_i)$ received in cyclic shear tests when initial normal pressure σ_{n_o} equalled: a) 54 kPa; b) 108 kPa; c) 216 kPa.

The second and main reason is the imperfect design of modern shear apparatuses; it concerns the influence of sample friction against the walls of the box. As a result, the shear band 'receives' a pressure which is lower than that applied to the surface of the sample. The error grows in proportion to the increase of the initial normal pressure (Amsheyus & Kuleshyus, 1982).

The first reason reflects objective physical properties of grain materials. Therefore, when precise measurements of strength are needed it is recommended to conduct tests within a definite range of pressures and to measure respective values of strength parameters therein (Fig. 4.24).

The methodical error can be eliminated by direct measuring of normal pressure in the shear band, which possibility is provided by the design of the contact friction apparatus DACS (see Section 3.2)

It is also necessary to mention conditions giving rise to the so-called strength of gear C_g measured in standard tests with coarse dense sands by the free dilatancy method, when the straight line $\tau_u = f(\sigma_{n_o})$ cuts off a section on the τ_u axis. The reason for the emergence of C_g is either the thrust of the sample along the walls of the box resulting in arch effect and a certain thrust in the shear band, or the same suppression of dilatancy (Fig. 1.10).

4.12 CYCLIC SHEAR IN CONDITIONS OF CONSTRAINED DILATANCY

The DACS-2 modification of the contact shear apparatus provides a possibility for cyclic application of shear force (see Section 3.2). A special series of tests was conducted with the help of this apparatus.

The testing of control-mix sands showed that cyclic shear appears to 'rock' the grain structure. The charts of function $\tau_i = f(\pm S_i)$ characterizing the manifestation of this effect are represented in Fig. 4.25. Upon reaching ultimate resistance to shear, the first reverse movement of the carriage as a rule mobilised only 70 to 80% of the initial value. At the same time the sample volume did not return to the initial, but stabilized at a constant level in the order of 65-70% of the initial, as well. Complete maximum-to-maximum shear strain constituted approximately 1.75 of the ultimate strain in respect to 'one-way' shear.

CHAPTER 5

Direct shear in conditions of constrained dilatancy

5.1 GENERAL PROPOSITIONS

This chapter highlights research conducted with the use of dilatometric apparatus of direct shear DADS and modernized standard shear apparatus BCB-25 employing the constrained shear method (Sections 3.3, 3.4). The experiments were aimed at establishing the peculiarities of 'soil-against-soil' shear, comparing the obtained results with the contact shear data, and matching experimental performances of the DACS, DADS and BCB-25 instruments.

In the course of tests the same control sand mixtures as in the case of DACS instrument were used (see Section 4.1). Analogous testing was carried out for sand samples with the density I_D = 1.0; 0.8; 0.6; 0.3.

During DADS instrument testing we attempted to model shear conditions both for equamodular soil medium, and for the contact of layers with different values of modulus of elasticity. In the former case the same rigidity values were set on dynamometers at both ends of the sample, while in the latter case these values were different.

Modelling the elastic reaction of the massif to dilatancy was ensured in these experiments by standard calibrating dynamometers and by regulated dynamometers. For the former, the rigidity for equamodular medium corresponded to values of coefficient of elastic resistance equalling K = 42, 1095 and 2644 MN/m³, whereas in the latter case (varimodular medium) the ratio of respective values of rigidity for the upper and the lower dynamometers ensured the given value of coefficient K_{red} = 1548 MN/m³ (see Section 2.3). Only dense sands were tested in the BCB-25 instrument (I_D = 1.0), and the modelled values of K_{red} equalled K_{red} = 83, 2189 and 5288 MN/m³, which corresponded to the values of rigidity of calibrating dynamometers. Additional experiments were conducted for intermediary values of K modelled by a regulated dynamometer with random-composition sands of varying degrees of density.

76 Direct shear in conditions of constrained dilatancy

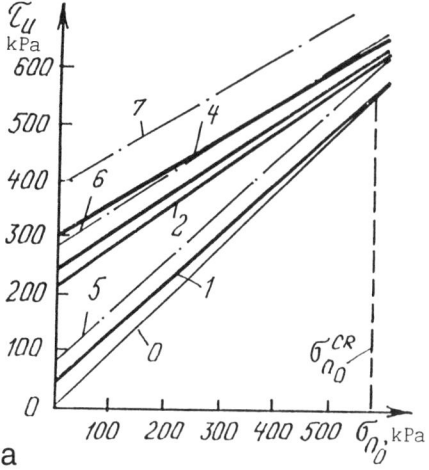

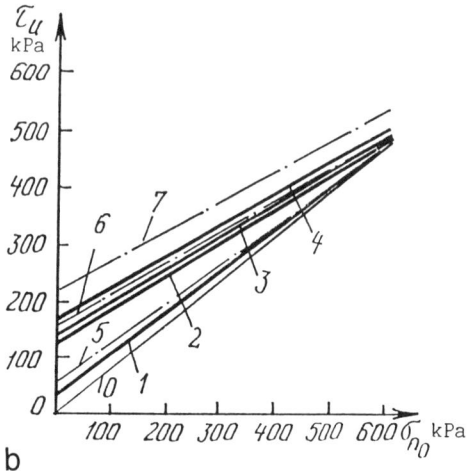

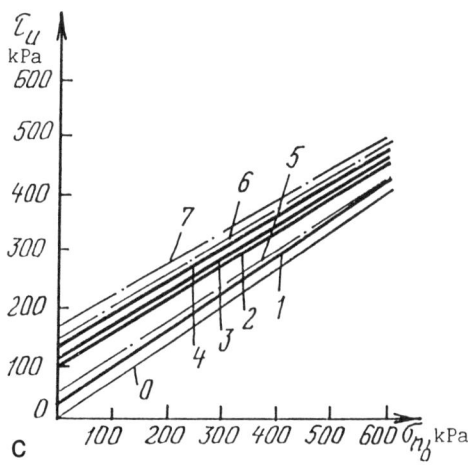

Figure 5.1. Functions 'ultimate resistance to shear τ_u – initial normal pressure σ_{n_0}' (according to the test data obtained with the use of DADS and BCB-25 instruments) a) coarse sand; b) medium sand; c) fine sand. 0. free dilatancy; 1. $K = 42$; 2. $K = 1095$; 3. $K = 1548$; 4. $K = 2644$; 5. $K = 83$; 6. $K = 2189$; 7. $K = 5288$ MN/m^3. DADS, –.– BCB-25.

5.2 INFLUENCE OF CONSTRAINT ON DILATANCY ON RESISTANCE TO SHEAR

The tests in the DADS and BCB-25 instruments conducted by the constrained dilatancy method corroborated the data obtained in the course of DACS experiments (see Section 4.2).

Fig. 5.1a, b, c represent experimental functions $\tau_u = f(\sigma_{n_o})$ for control-mix sands of dense composition. One can easily observe the above-mentioned pattern where an increase of elastic resistance to dilatancy is accompanied by an increase of the dilatant component of strength τ_d and a decrease of the angle of internal friction φ'.

The same regularity can be traced in experimental data Tables 5.1 and 5.2 for sands of varying density and grain size. The character of deformation of samples in DADS and BCB-25 tests was practically the same as in the above-mentioned case of the DACS instrument. The only difference is the wider spread of the obtained values, especially during tests on samples with values of density equalling $I_D \leq 0.6$, which demanded a larger number of tests. The reasons for this phenomenon will be supplied later.

So, in the case of the 'soil-against-soil' shear the Coulomb function is generally analogous to (4.1):

Table 5.1. Experimental values of dilatant stress σ_d, kPa dilatant component of strength τ_d, kPa and angles of internal friction φ, grades for quartz dry sands according to data obtained with the help of DADS instrument.

Sands under study	Given values of coefficient of elastic resistance K, MN/m³	Index of density I_D											
		1.0			0.8			0.6			0.3		
		σ_d	τ_d	φ'	σ_d	τ_d	φ'	σ_d	τ_d	φ'	σ_d	τ_d	φ'
Coarse $d_{50}=1.50$ mm	0	0	0	44.1	0	0	43.0	0	0	41.0	0	0	40.4
	42	44	40	42.0	48	44	42.6	36	31	40.6	10	8	39.0
	1095	297	208	35.0	210	147	35.0	145	100	34.5	45	26	30.0
	1548	347	230	33.5	252	164	33.0	177	111	32.2	51	29	29.5
	2644	471	292	31.5	343	199	30.0	238	138	30.6	76	37	26.1
Medium $d_{50}=0.50$ mm	0	0	0	38.0	0	0	36.0	0	0	35.3	0	0	34.0
	42	33	24	35.7	44	31	35.1	25	17	35.1	4	3	34.0
	1095	208	125	31.0	103	63	31.4	66	40	31.3	16	9	28.8
	1548	241	140	30.0	137	81	30.6	83	48	30.3	21	11	27.9
	2644	286	155	28.5	162	89	29.0	–	–	–	28	13	24.0
Fine $d_{50}=0.25$ mm	0	0	0	35.0	0	0	35.0	0	0	34.3	0	0	34.0
	42	29	20	35.0	34	21	34.7	20	13	34.1	0	0	34.0
	1094	157	96	31.4	109	63	30.0	69	40	30.0	10	6	29.8
	1548	172	100	30.3	128	72	29.2	77	43	29.3	11	6	27.8
	2644	221	124	29.3	157	85	28.5	93	50	28.3	21	10	25.0

Table 5.2. Experimental values of dilatant stress σ_d, kPa dilatant component of strength τ_d, kPa and angles of internal friction φ, grades for quartz dry dense sands according to the data obtained with the help of the BCB-25 instrument.

Sands under study	Values of coefficient of elastic resistance K, MN/m^3	Index of density $I_D = 1.0$		
		σ_d	τ_d	φ'
Coarse $d_{50} = 1.50$ mm	0	0	0	44.1
	83	78	72	42.8
	2189	435	283	33.0
	5288	635	383	31.0
Medium $d_{50} = 0.50$ mm	0	0	0	38.0
	83	66	48	35.8
	2189	274	153	29.3
	5288	399	212	28.1
Fine $d_{50} = 0.25$ mm	0	0	0	35.0
	83	56	37	32.9
	2189	221	128	29.7
	5288	294	159	28.2

$$\tau_u = \tau_d + \sigma_{n_o} \operatorname{tg} \varphi' = (\sigma_d + \sigma_{n_o}) \operatorname{tg} \varphi', \qquad (5.1)$$

where φ' is the angle of internal friction. When resistance to dilatancy is removed, τ_d turns into zero, while φ' reaches its maximum value φ_0; which corresponds to the case of full realization of the component of gear angle (see Sections 1.2 and 4.4).

The values of critical normal initial pressure (complete constraint of dilatancy) are given in Table 5.3. Here, as in the case of contact shear (Table 4.3), we observe a tendency for the pressure $\sigma_{n_o}^{cr}$ to lower parallel to the increase of massif's rigidity.

The obtained results were substantially influenced by the clearance between rings of the box of the movable and immovable carriages, the slant of pistons and the possible loosening of the sample occurring in the process of its placement into the instrument. Empirically the optimum value of clearance between rings in the shear plane was established to be equal to 2.0-$2.5 d_{50}$.

5.3 DILATANT COMPONENT OF STRENGTH AND ANGLE OF INTERNAL FRICTION AS FUNCTIONS OF MASSIF'S ELASTICITY

Fig. 5.2 represents functions $\tau_d = f(K)$ obtained in the course of tests with all the three shear instruments (DACS, DADS, BCB-25). It can be seen in the charts that during the 'soil-against-soil' shear we observe the same parabolic pattern of the equation (4.5), as in the case of contact shear, but the obtained values are slightly higher than those measured in the DACS instrument.

This difference can be easily accounted for. While with contact shear the sliding surface is traced beforehand, in the process of direct shear additional work is required to form it, to be more specific, to overcome stronger grain gear.

This is confirmed by the data presented in Table 5.4, which gives coefficients a and α for different types of tests (see Sections 4.3-4.5). The largest divergency of values of a can be observed for coarse sand and for maximum-density medium and fine sand. In other words, the gear is in direct proportion to the grain size and density of grain packing.

In their turn, the values of τ_d measured during tests in the BCB-25 instrument are slightly higher than those in the case of the DACS instrument. This is connected with the presence of conditions for larger constraint of dilatancy. If in the DADS instrument dilatancy can develop in the directions of both ends of the sample, in the BCB-25 instrument it is only one direction. The errors resulting

Table 5.3. Experimental values of critical initial normal pressure $\sigma_{n_o}^{Cr}$, kPa (complete suppression of dilatancy) according to the data for quartz sands tested in the DADS and BCB-25 instruments.

Sands under study	Instrument	Given values of coefficient K (MN/m³)	Index of density I_D		
			1.0	0.8	0.6
Coarse $d_{50} = 1.50$ mm	DADS	42	1883	1392	–
		1095	773	634	408
		1548	749	579	366
		2644	719	561	305
	BCB-25	83	1672		
		2189	1046		
		5288	885		
Medium $d_{50} = 0.50$ mm	DADS	42	1680	1312	–
		1095	694	580	400
		1548	686	567	390
		2644	651	517	–
	BCB-25	83	795		
		2189	856		
		5288	695		
Fine $d_{50} = 0.25$ mm	DADS	42	–	–	
		1095	1068	552	
		1548	862	541	
		2644	842	508	
	BCB-25	83	994		
		2189	794		
		5288	669		

* Note: Tests were conducted with the use of standard calibrating dynamometers with the sample section area of 40 square cm.

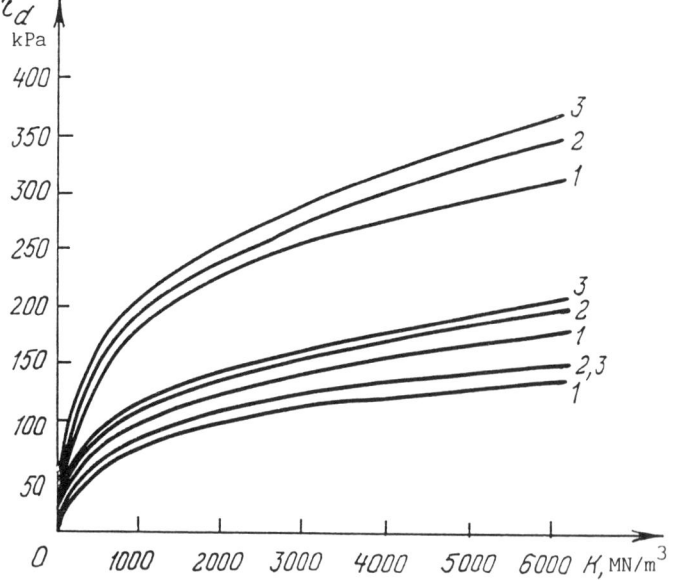

Figure 5.2. Functions $\tau_u = f(K)$ for dense sands: coarse sands (the top curves); medium sands (the middle curves) and fine sands (the lower curves) according to test data obtrained with the use of the instruments: 1. DACS; 2. DADS; 3. BCB-25.

from this fact were not large, though, not exceeding 10%.

The functions $\varphi' = f(K)$ for coarse sand are represented in Fig. 5.3. For the angle of internal friction φ' during the 'soil-against-soil' shear in the conditions of constrained dilatancy, the empirical function (4.9) deduced from contact shear tests proves equally correct. We also observe a decrease of the angle of internal friction φ' with the increase of massif's rigidity in the direction of the value of the angle of intergranular friction φ_μ. But the values of φ' are somewhat higher than those of φ'_c, which can be apparently accounted for by a larger component of gear $\varphi_g = \varphi_o - \varphi_\mu$. When the grain size and density of sands decrease, the above-noted difference between the angles of internal and contact friction disappears. This is manifested by the same values of coefficient α for different tests (see Table 5.4).

A comparison of data received during tests with the DADS and BCB-25 instruments reveals that the latter yields slightly overread values of the τ_d and φ' parameters. But considering the insignificance of this error, we deem its employment in constrained dilatancy tests quite acceptable.

5.4 COMPARISON AND PECULIARITIES OF TESTS IN DILATOMETRIC INSTRUMENTS OF VARIOUS DESIGNS

Tests conducted with the DADS and BCB-25 instruments did not generally reveal any substantial differences in sand behaviour from the case of the contact shear. Among the peculiarities we could list somewhat greater values of resistance to shear for coarse sands, and, consequently, larger values of the parameters τ_d and φ'. This is vividly demonstrated in Table 5.4 while comparing empirical coefficients a and α.

With the decrease of sand density we can also observe a drop in the values of resistance to shear (see Tables 5.1 and 5.2). A detailed analysis of the density

Table 5.4. Values of empirical coefficients a and α according to contact shear test data (the DACS instrument) and direct shear test data (the DADS and BCB-25 instruments).

Sands under study	Test type	Instrument	Values of $a \times 10^{-4}$ (MN/m²)$^{2/3}$ and $\alpha \times 10^{-4}$ (m³/MN)							
			1.0		0.8		0.6		0.3	
			a	α	a	α	a	α	a	α
Coarse $d_{50} = 1.50$ mm	Contact shear Direct shear	DACS DADS BCB-25	185 208 220	8.0 6.1 4.8	138 142 –	7.3 7.3 –	94 97 –	6.5 6.5 –	28 – –	– – –
Medium $d_{50} = 0.50$ mm	Contact shear Direct shear	DACS DADS BCB-25	106 117 118	12.0 11.5 5.8	62 62 –	8.7 8.7 –	40 40 –	7.6 7.6 –	9 – –	– – –
Fine $d_{50} = 0.25$ mm	Contact shear Direct shear	DACS DADS BCB-25	84 88 88	13.0 13.0 9.0	60 60 –	12.0 12.0 –	37 37 –	11.0 11.0 –	5.4 – –	– – –

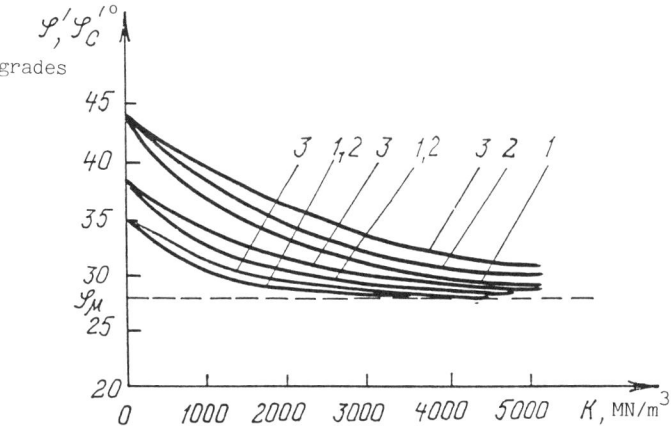

Figure 5.3. Functions φ', $\varphi'_c = f(K)$ for dense sands: coarse sands (the top curves); medium sands (the middle curves); fine sands (the lower curves) according to test data obrained with the use of the instruments: 1. DACS; 2. DADS; 3. BCB-25.

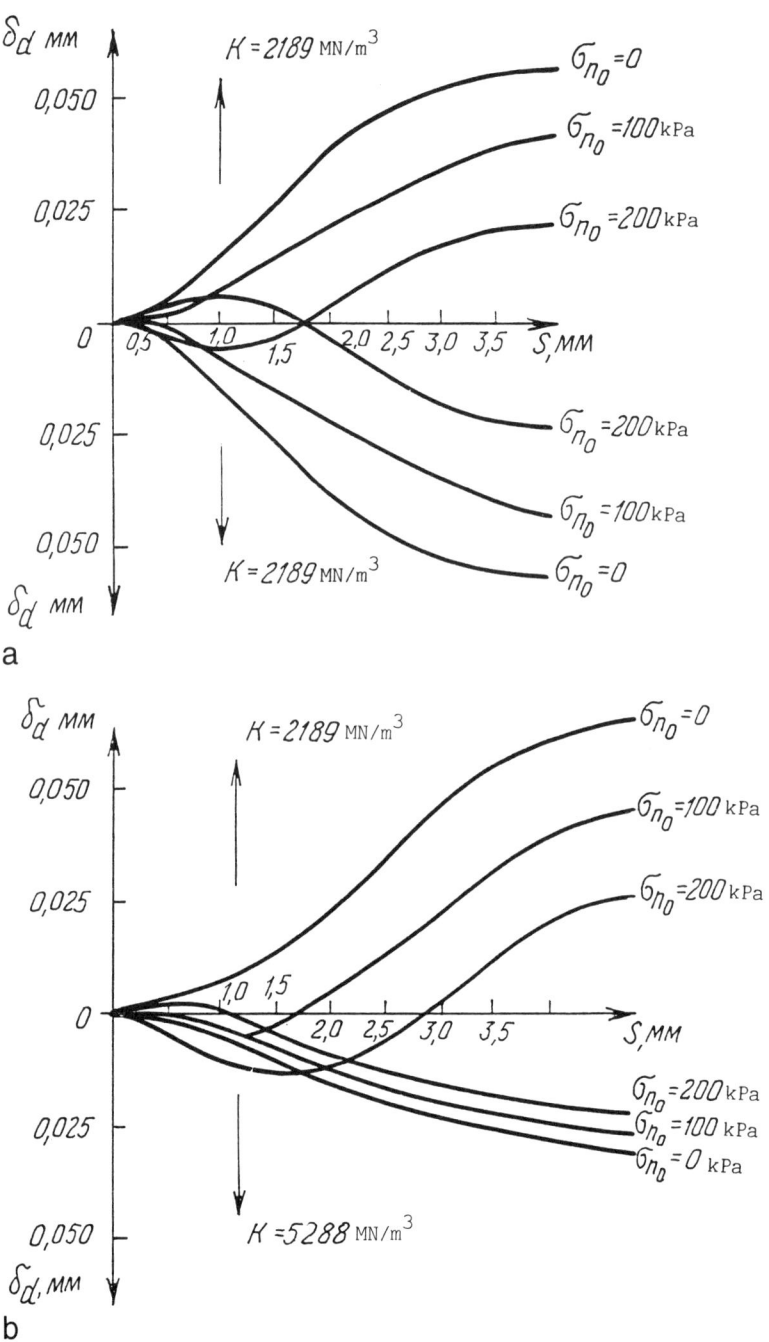

Figure 5.4. Functions $\delta_{d_i} = f(S_i)$ for coarse sand when the process of shear is modelled: a) in equamodular medium ($K = 1095$ MN/m³); b) in varimodular medium ($K_{red} = 2644$ MN/m³).

factor is available in Section 4.5. A decrease of initial density results in a lower dilatant component τ_d due to a larger freedom of grain repacking. For the angle φ' this is tantamount to a lower gear component.

In Fig. 5.4 there are several functions $\delta_d = f(S_i)$ for coarse sands according to test data obtained from experiments with the use of DADS instrument. Due to the fact that in this instrument the sample can dilate in the directions of both its ends dilatancy is characterized by two sets of curves. They are positioned symmetrically to the abscissa when the elastic connection is characterized by equal values of rigidity both on top and at the bottom of the sample (Fig. 5.4a), and asymmetrically when rigidities are different (Fig. 5.4b). Otherwise the functions $\delta_d = f(S_i)$ are analogous to those obtained for the contact shear case (Fig. 4.22).

At the initial stages of the shear with $\sigma_{n_o} = 100\text{-}150$ kPa the curves form initial loops of contraction, which reflects the period of mobilization of grain gear. At certain initial critical pressure $\sigma_{n_o}^{cr}$ both the branches of curves converge to the axis S_i, which signifies total suppression of dilatancy. The exceeding of the initial normal pressure $\sigma_{n_o}^{cr}$ causes the sample to contract. Limiting movements corresponding to the exhaustion of resistance to shear constituted in these experiments 4.5-5.5 mm, as a rule.

A comparison of test data of contact shear and direct shear experiments has not revealed any differences in the influence exerted by humidity, the value of dilatant and limiting shear movements, either, therefore the concepts expounded in Chapter 4 can be fully applied to the present material. Onc has to bear in mind, though, that the failure plane is predetermined by the instrument's design, and while it perfectly corresponds to natural conditions in the case of contact shear, with 'soil-against-soil' shear a considerable amount of work is needed in order to form the sliding surface. Besides, in a DADS apparatus the shear surface acquires a lens form. As for dilatant movements and thrust stresses, they are normal to it.

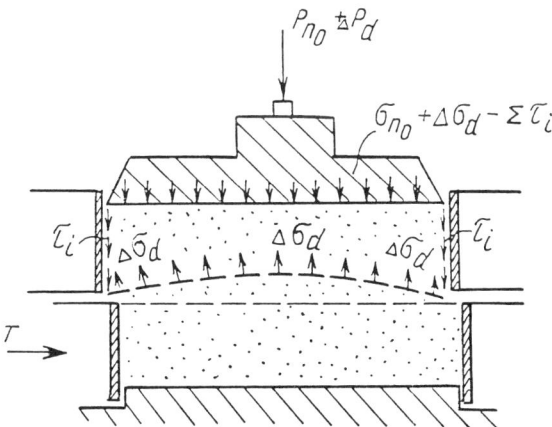

Figure 5.5. Shear surface in a sample. Distribution of dilatant thrust σ_{di} and friction along box walls τ_i.

Correspondingly, a larger share of these stresses goes to the mobilization of additional friction along the sample-box walls (Fig. 5.5).

5.5 CONCLUSION

Constraint of dilatancy leads to changes in the soil's stress condition in the process of shear as a result of the thrust emerging with grain repacking in the sliding surface. The appearing dilatant stresses are in direct dependence to the rigidity of the massif, the soil density, the grain size and the strength of the forming mineral. When shear occurs in the soil with density not exceeding the initial critical value, the gear partly disappears, there is contraction with an unloading of initial stresses, resistance to shear diminishes. Constrain of dilatancy changes the very character of dependence of resistance to shear for non-cohesive soils. Unwedging of sliding surface grains causes the emergence of the dilatant component of resistance to shear, and the value of this component is determined by the massif's rigidity. The increase of rigidity makes the shear mechanism similar to the condition of mutual sliding of solid elastic blocks. The angle of internal (contact) friction with free dilatancy decreases in the direction of the value of the angle of intergranular friction φ_μ and acquires a certain intermediary value $\varphi'_{(c)}$. Correspondingly, Coulomb's law take the equations (4.1), (4.2) and (5.1).

The dilatant component of strength τ_d and the angle of internal (contact) friction $\varphi'_{(c)}$ as functions of deformative parameters of soil massif can be quite satisfactorily approximated by empirical functions of the equations (4.7), (4.8) and (4.10), (4.11). The exclusion of dilatancy constraint conditions returns Coulomb's law to its original (traditional) form, i.e. when $K = 0$, $\tau_d = 0$ and $\varphi'_{(c)} = \varphi_o$. Consequently, Coulomb's law for non-cohesive media in its traditional reading as $\tau_u = \sigma_{n_o} \operatorname{tg} \varphi_o$ is true only for free dilatancy conditions and represents a specific case of equations (4.1) and (5.1).

Traditional laboratory test methods where the sample volume can freely change in the process of failure give erroneous values of resistance to shear when resistance to dilatancy is present in the soil. A coincidence of values of τ_u is possible only at critical initial normal pressure. Similarly, the use of instruments which merely restrict the freedom of volume dilating of the sample creates uncertainty in the conditions determining resistance to shear.

The following directions for further research might be suggested:
1. Refinement of the empirical functions (4.6) and (4.9) by means of:
    a) Taking into account not only the mean grain diameter, but the whole range of these values, namely d_{10}, d_{20}, d_{30} etc.
    b) Additional study of the influence exerted by humidity, especially in the case when silty and clayey fractions are represented in the soil composition;
    c) Redetermination of the influence of grain strength in various pressure ranges etc.

It would be justified to allow for these and other similar factors by dividing coefficients a and α into several coefficients, each of these latter being 'responsible' for its own factor determining soil behaviour, for instance,

$$\tau_d = a_1 a_2 \text{-} a_n K^n \tag{5.2}$$

$$\varphi'_{(c)} = \varphi_\mu + (\varphi_o - \varphi_\mu)\exp - \alpha_{1,2}\text{-}\alpha_n K \tag{5.3}$$

where a_1, α_1 are coefficients reflecting the influence of initial density of packing; $a_2\text{-}a_n$, $\alpha_2\text{-}\alpha_n$ – are coefficients reflecting the influence exerted by various factors as described above in a), b), c).

2. Spreading the research on the constrained dilatancy lines to the shear of coarse fractured soils, pebbles, stone fills and other kinds of grainy media.

3. Using the method set forth herein for a study of clayey soils, which contain substantial (in terms of volume) fractions of sand and coarse-grained fractions, for example, overcompacted (glacial) soils.

When a coarse-grain skeleton is present in such soils, a considerable dilatancy is not unlikely. If this dilatancy is subjected to constraint, it causes great changes in shear strength. Tentative experiments conducted by the author on hard moraine loam with the use of DADS and DACS seem to prove this point. In a number of cases it would possibly be justified to write Coulomb's law in the equation

$$\tau_u = C + \tau_d + \sigma_{n_o} \text{tg } \varphi'_{(c)} \tag{5.4}$$

where C is specific cohesion.

4. Conducting special research on constant-speed shear methods, and refinement in connection with this of functions of the type $\tau_i = f(S_i)$ in order to allow for peak and residual strength.

5. Conducting research of contact shear with deforming of the shear surface. Such research which is beyond the scope of this book, is possible with the use of instruments of the DACS-A type (see Section 3.5).

6. Studying soil behaviour in the conditions of multiple cyclic application of shear force and the influence of this behaviour on the increment of stresses due to dilatancy.

CHAPTER 6

Internal bulge as a manifestation of conditions of constrained dilatancy

6.1 GENERAL PROPOSITIONS

Resistance to internal bulge determines the bearing capacity of soil under the lower end of piles, trench foundations and other types of deep supports. Internal bulge is characterized by the absence of conditions of free dilating of plastic deformation area. According the accepted model, dilatancy occurring all throughout this area develops with elastic resistance of the surrounding massif (see Section 2.4).

The conditions of internal bulge are best modelled in triaxial compression instruments during sample-squashing tests: $\sigma_1 > (\sigma_2 = \sigma_3)$. Such tests are usually conducted with a gradual increase of vertical pressure relatively to lateral pressure which is kept constant throughout the experiment (see Section 1.7). With this method the dilatancy of the sample does not influence its stress condition, but only causes the flowing of liquid out of the camera of the instrument. In other words, the modelled conditions of free dilatancy are similar to those reproduced in shear instruments of traditional design.

The research described later aimed at taking into account the influence of constraint of volume deformations on the soil behaviour and soil strength in the conditions of internal bulge. The tests were conducted with the use of dilatometric triaxial apparatus (DTA) (see Section 3.6) on cylindrical samples of low-humid sands of control mixes (see Section 4.1) with a cross section area of 25 cm² and a height of 11-13 cm. The loading was done according to the sample-squashing scheme. In order to do this, after having created hydrostatic pressure on the sample, the vertical pressure was increased. At each loading stage we waited till the deformations stabilized (not less than 5 minutes) and then registered the increment of lateral pressure $\Delta\sigma_d$ and the piston travel $\Delta\delta_d$. These values depended on the rigidity of the dynamometer which was used to model the massif's reaction to dilatancy (see Section 2.4).

Standard calibrating dynamometers were used during the tests. The dynamometers modelled values of $K = 70$, 1933 and 4552 MN/m³ with the lateral piston diameter equal to 56.4 mm. Control tests were conducted with intermediary values of K set on a regulated dynamometer. Thus, soil strength properties were defined in correlation with the modelled deformative characteristics of the massif.

In the text below, σ_1 and σ_3 are used to designate respectively the highest and the lowest principal stresses (vertical and lateral) mobilized at the moment of failure during constrained dilatancy tests. While describing free dilatancy tests, the same stresses are designated as σ_{01} and σ_{03}; σ_{03} is also used to designate initial lateral pressure on the sample.

6.2 INFLUENCE OF DILATANCY ON STRESS-DEFORMATIVE CONDITION OF SOIL DURING TRIAXIAL COMPRESSION

This influence can be evaluated by considering the functional relationship between the principal stresses σ_1 and σ_3 on the one hand, and the values of coefficient of elastic resistance K on the other hand. Tables A1.4 and A1.5 give the values of σ_1 and σ_3, angles of internal friction φ' and volume increments ΔV_d (dilatancy) for sands of varying grain size and density tested by both free and constrained dilatancy methods. A comparison of these tables yields a substantial difference between the mobilized values of principal stresses, their ratios and strength parameters.

Regularities close to those established during shear-instrument testing of the same sands (see Chapters 4, 5) are observed. An increase of elastic resistance to dilatancy leads to an increase of mobilized values σ_1 and σ_3 with a simultaneous decrease of the angle of internal friction φ'. Fig. 6.1 represent charts of the functions 'principal stresses-coefficient of elastic resistance K' with different values of initial (hydrostatic) pressure on the sample ($\sigma_{01} = \sigma_{03}$).

A constraint of dilatancy results in the fact that an increase of lateral pressure on the sample simultaneously causes mobilization of additional resistance to its squashing. Consequently, soil strength during modelling of internal bulge conditions turns out considerably higher, than the strength measured in triaxial tests with constant lateral pressure. This opinion can be substantiated by comparing the curves $\sigma_{1,3} = f(K, \sigma_{01}, \sigma_{01})$ with the lines $\sigma_1 = f(\sigma_{03} = \text{const})$ in Fig. 6.1. The difference of the ordinates gives the value of stress increment due to dilatancy. Here we can observe the same parabolical functions $\sigma_d = f(K)$, as those established earlier in shear tests (Fig. 5.2).

6.3 ADJUSTMENTS FOR DILATANCY IN COULOMB-MOHR'S STRENGTH CONDITIONS

The Coulomb-Mohr strength condition for non-cohesive soil with free dilatancy takes the form

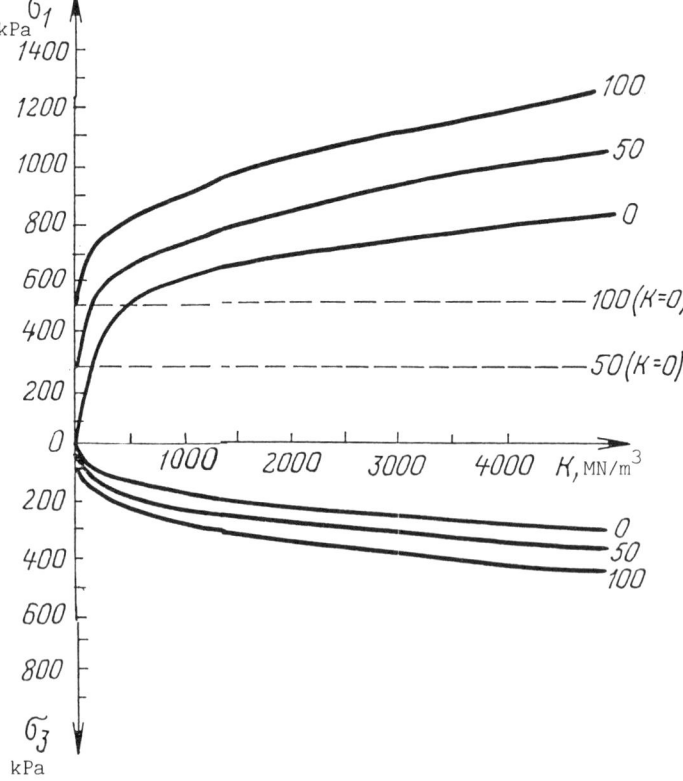

Figure 6.1. Functional relationship of principal stresses σ_1, σ_3 to the modelled coefficient of elastic resistance for coarse dense sand with initial lateral pressure on the sample $\sigma_{03} = 0; 50; 100$ kPa.

$$\frac{\sigma_{01} - \sigma_{03}}{\sigma_{01} + \sigma_{03}} \leq \sin\varphi_o \qquad (6.1)$$

With constraint imposed on dilatancy during sample-squashing tests the lateral pressure becomes a function of coefficient of elastic resistance, i.e. $\sigma_3 = f(K)$. The lateral pressure receives an increment by the value of the mobilized dilatant stress, i.e.

$$\sigma_3 = \sigma_{03} + \Delta\sigma_d \qquad (6.2)$$

A corresponding increment is received by the vertical pressure. In this case the ratio between the principal stresses becomes dependent on the degree of constrain of volume deformations, which is reflected by the changes in the measured value of the angle of internal friction φ'. This is clearly seen if one compares the tangent lines to the Mohr circle for free and constrained dilatancy tests (Fig. 6.2). This figure also illustrates the hydrostatic action of dilatant stresses. Generally the condition of strength of dilating soil can be expressed as

$$\frac{\sigma_1 - \sigma_3}{\sigma_1 + \sigma_3 + 2\sigma_d} \leq \sin \varphi' \qquad (6.3)$$

When $\sigma_d = 0$ the condition (6.3) takes the equation (6.1) which, consequently, is a specific case of the former.

If only the initial value of lateral pressure σ_{03} and the coefficient of elastic resistance are known, the resistance of soil to internal bulge can be expressed by equation (6.3) as follows:

$$\sigma_1 = \sigma_{03} \, \text{tg}^2 (45° + \frac{\varphi'}{2}) + 2\sigma_d \, \text{tg} \, \varphi' \, \text{tg} (45° + \frac{\varphi'}{2})$$

$$= \sigma_{03} \, \text{tg}^2 (45° + \frac{\varphi'}{2}) + 2\tau_d \, \text{tg} (45° + \frac{\varphi'}{2}) \qquad (6.4)$$

Let it be remembered that here, σ_{03} designates initial lateral pressure along the contour of the failure area, and that the values of τ_d in this case also follow functions of the equation (4.6).

Fig. 6.3 represents a sample-squashing test in stress-condition invariants, which can be expressed
– for the case of free dilatancy ($K = 0$, $\sigma_{03} = \text{const}$) as

$$T_{pl_0} = \frac{\sigma_{01} - \sigma_{03}}{2} \text{ and } \sigma_{pl_0} = \frac{\sigma_{01} + \sigma_{03}}{2}, \qquad (6.5)$$

– for the case of constrained dilatancy ($K > 0$, $\sigma_{3,1} = f(K)$) as

$$T_{pl} = \frac{\sigma_1 - \sigma_3}{2} \text{ and } \sigma_{pl} = \frac{\sigma_1 + \sigma_3}{2} \qquad (6.6)$$

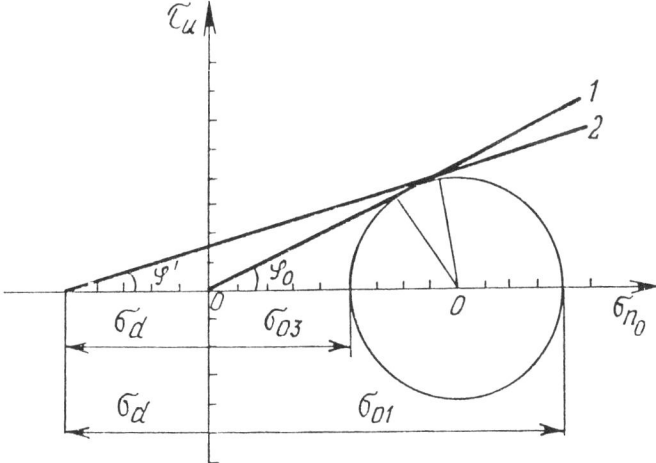

Figure 6.2. Tangent lines to Mohr circle for the cases of: 1. free dilatancy; 2. constrained dilatancy.

90 *Internal bulge as a manifestation of constrained dilatancy*

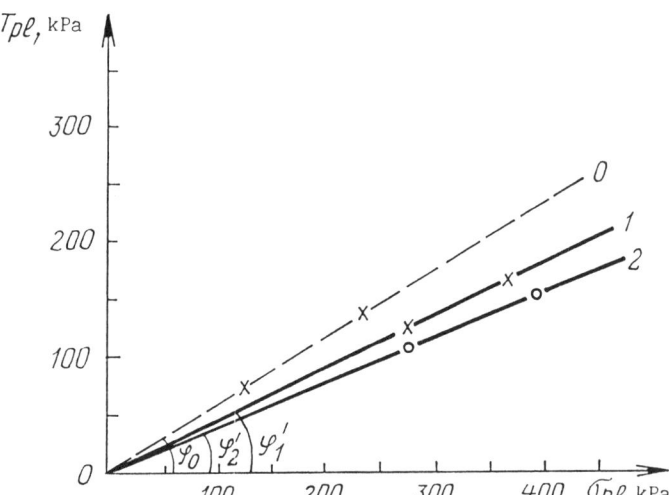

Figure 6.3. Presentation of sample-squashing test in invariants of stress condition: 0. with free dilatancy (σ_{03} = const); 1. with K = 1933 MN/m³; 2. with K = 4552 MN/m³.

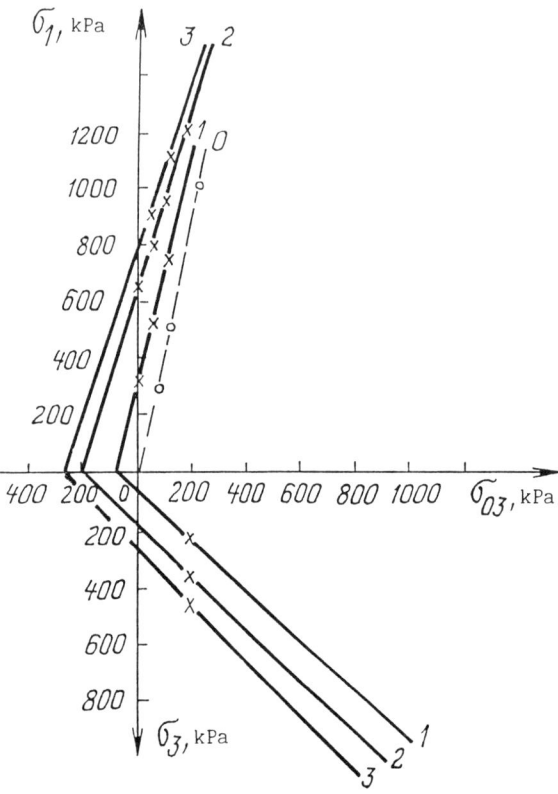

Figure 6.4. Functional relationship of principal stresses σ_1 and σ_3 to initial lateral pressure σ_{03} for coarse dense sand. 0. with free dilatancy; 1. with K = 70 MN/m³; 2. with K = 1933 MN/m³.

According to (6.2) we can write down:

$$T_{pl} = \frac{\sigma_{01} - \sigma_{03}}{2} = T_{pl_0} \text{ and } \sigma_{pl} = \frac{\sigma_{01} + \sigma_{03}}{2} + 2\sigma_d = \sigma_{pl_0} + 2\sigma_d \quad (6.7)$$

The given chart has the conventional form for representing the triaxial test results in the same coordinates regardless of the presence or absence of constraint on dilatancy. From this chart it is clear that the angle of internal friction decreases with the increase of K, but it does not indicate to what degree it is compensated for by the growth of internal thrust stresses due to dilatancy.

The chart in Fig. 6.4 represents the dependence of mobilized principal stresses on the initial lateral σ_{03} in absolute values. It enables us to judge the linear increase of resistance to squashing with the growth of σ_{03}. The straight lines $\sigma_{1,3} = f(\sigma_{03})$ do not begin at the origin of coordinates; when prolonged until their intersection with the abscissa they converge, laying off the values of dilatant stress σ_d, which corresponds to the value of coefficient of elastic resistance K. We can observe the same regularities as for the the functions (4.1) and (5.1) set forth above.

6.4 ANGLE OF INTERNAL FRICTION AS A FUNCTION OF MASSIF ELASTICITY

The change in the ratios of principal stresses with elastic constraint to dilatancy leads to a decrease in the measured values of the angle of internal friction. The data given in Table A1.4 testify to the fact that when $K = 4552$ MN/m^3, the values of φ' are close to the values of the angle of intergranular friction $\varphi_\mu = 28°$ (see Section 1.2).

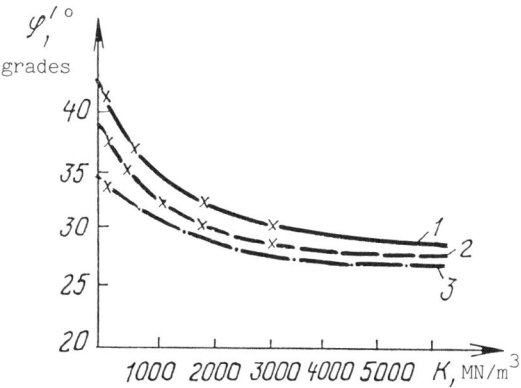

Figure 6.5. Functions $\varphi' = f(K)$ for sands: 1. coarse; 2. medium; 3. fine.

Fig. 6.5 represents functions $\varphi' = f(K)$ for medium sands. They are completely identical to the functions in Figs 4.6 and 5.3 and are also described by an empirical function of the equation (4.9). The values of φ' for control-mix sands measured in the course of triaxial and shear experiments diverge very slightly if at all.

The question then arises regarding the reasons for the decrease of φ' with the constraint of dilatancy. If in the case of shear it is rather easy to present this as an approximation of the sliding mechanism to the mutual movement of solid blocks, in the case now under consideration the failure spreads over a large gear, involving a correspondingly large number of grains into repacking. Probably the answer to this question should be sought in the mechanism of achieving the condition of critical density. Suppression of dilatancy limits possibilities for volume deformation and, consequently, for mutual grain movement; and the grain movement taking place by the moment, when the limiting ratio between the principal stresses σ_1 and σ_3 is achieved, is sufficient only for realization of the angle of intergranular friction and, partially, gear.

6.5 PECULIARITIES OF SOIL DEFORMATION WITH CONSTRAINED DILATANCY

Deformation of samples during constrained dilatancy tests always took the form of barrel-like billowing. Parallel to the increase of vertical pressure this billowing grew more pronounced together with the growth of the sample volume. This was accompanied by displacement of liquid from the camera, which was registered by the lateral piston travel (Fig. 3.9). Upon reaching the ultimate vertical pressure σ_v, the lateral pressure σ_h remained constant and further deformation of the sample went without volume changes, i.e. solely at the expense of form changing (Fig. 6.6). This fact proves that in the case of internal bulge ultimate strength corresponds to the acquisition of the critical density condition by the soil in the failure area.

The experiments revealed that the values of sample volume deformations ΔV_d are a function of the coefficient of massif elastic resistance K (see Table A1.5). At the same time the influence of initial lateral stress σ_{03} is not very significant. Partial suppression of dilatancy (approximately by 20-30%) occurred only with $\sigma_{03} = 200$ kPa, and this fact can be accounted for by strong preliminary soil compaction.

The functions $\Delta V_d = f(K)$ in Fig. 6.7 have an exponential outline with an especially sharp drop of dilatancy in the range of $K = 0...2000$ MN/m^3. At higher values of initial lateral pressure σ_{03} and of coefficient K, contraction of soil in the bulge area is possible due to a denser grain repacking or failure of grains at contact points. Ceteris paribus, larger dilatancy with correspondingly larger increments of principal stresses is observed for coarser and denser sands, which fact is illustrated by the chart in Fig. 6.8.

Peculiarities of soil deformation with constrained dilatancy 93

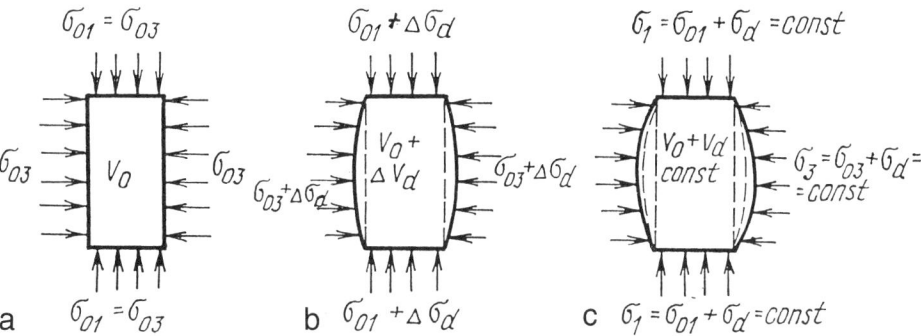

Figure 6.6. Stages of deformative – stress condition of the sample during a constrained dilatancy test: a) initial, consequent to volume compression and stabilization of deformations; b) intermediary, increase of stresses leads to form-changing deformations and increased volume (dilatancy); c) ultimate, the ratio of principal stresses is constant, deformations are solely form-changing.

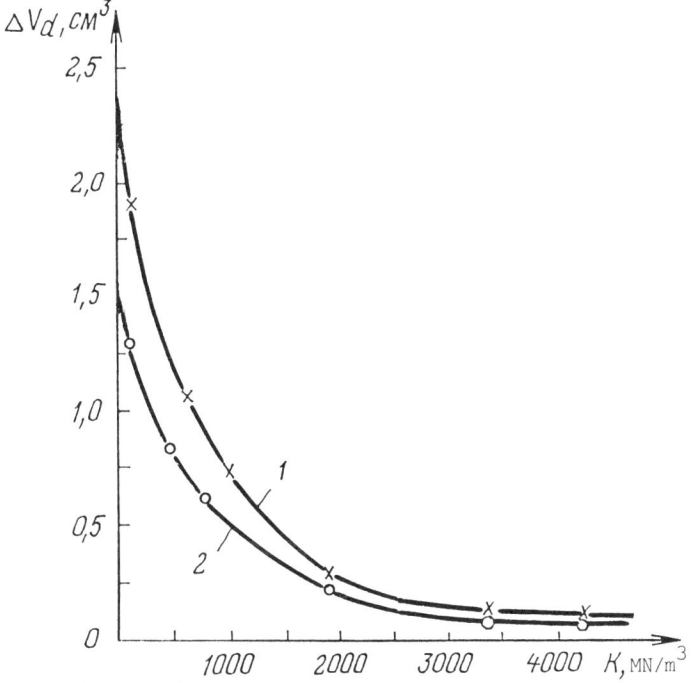

Figure 6.7. Functional relationship of volume deformations to coefficient of elastic resistance for coarse sand, when the initial density index I_D is equal to: 1. 0.8; 2. 0.6.

The chart of the function 'principal stress – initial packing density' is represented in Fig. 6.9. The obtained functions are well enough interpreted by straight lines, which can possibly be explained by the fact that during the experiments we used sands, received from the same original material by adding this or that fraction.

94 *Internal bulge as a manifestation of constrained dilatancy*

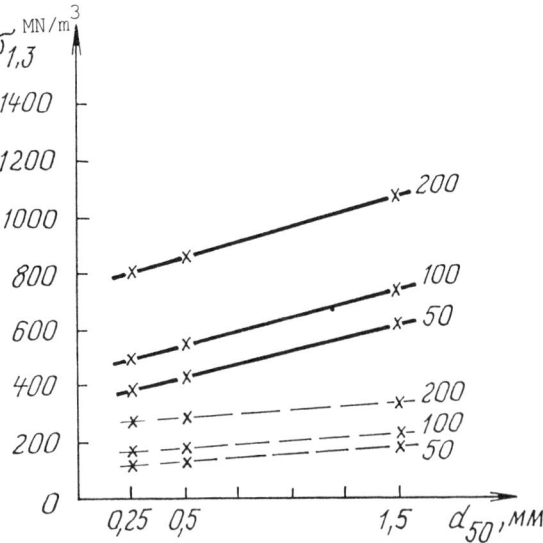

Figure 6.8. Influence of mean grain size of dense sand on the mobilized values of principal normal stresses with initial pressures σ_{03} = 50, 100 and 200 kPa, K = 1933 MN/m³ ——— σ_1; ---- σ_3.

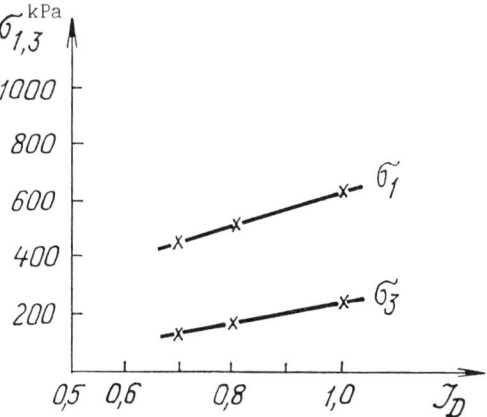

Figure 6.9. Influence of initial density index I_D on the mobilized values of principal normal stresses on the example of fine sand tested at σ_{03} = 100 kPa, and K = 1933 MN/m³.

6.6 CONCLUSION

Generalizing, we might conclude as follows: resistance of dilating soil is proportional to the mean size of the grains it consists of, and to the initial density of their packing. Thus, sand deformation regularities noted for shear and triaxial tests are identical.

Constraint of dilatancy is a reserve of additional soil strength during internal bulge, which is ignored by the currently accepted concepts. It leads to changes in the ratios between the principal stresses at the moment of failure; the value of the angle of internal friction decreases in the direction of the value of the angle of intergranular friction.

Maximum dilatancy in the process of internal bulge occurs at the moment of soil failure. With further loading resistance of soil remains constant, and its deformation takes place at the expense of form changing (when critical density of packing is achieved, the sample dilates into a barrel-like form). So, the condition of strength with constrained dilatancy takes the equation (6.3) and the maximum resistance to internal bulge can be represented for practical calculations in the equation (6.4) through the initial lateral pressure in the massif.

It is worth noting that traditionally-designed triaxial compression instruments and methods of testing with constant lateral pressure on the sample give considerable errors in determining the resistance to internal bulge and dilating soil strength parameters corresponding to this resistance.

The following directions for further research can be suggested:

1. Studying the mechanism of realization of the gear angle when soil failure occurs in a closed volume.

2. Conducting tests which model the process of unloading the plastic deformation area, including the specific case of cyclic application of the load.

3. Studying soil behaviour in the complex stress condition with dynamic application of the load.

4. Modelling relaxation process in the massif with prolonged loading of the sample (see Section 2.6).

5. Studying the influence of porous pressure during drainage and non-drainage soil tests, especially when the soil contains silty and clayey fractions.

6. Further research with the use of the constrained dilatancy method for triaxial stress condition as applied to coarse-grained soils, pebbles, other granular mediums.

7. Employing the triaxial test method for researching clayey soils containing substantial (in terms of volume) sand fraction and coarse-grained fraction (especially, overcompacted). The Rankine condition for similar soils could probably be represented in the following form:

$$\frac{\sigma_1 - \sigma_3}{\sigma_1 + \sigma_3 + 2(C + \tau_d)\,\mathrm{tg}\,\varphi'} \leq \sin\varphi' \qquad (6.8)$$

CHAPTER 7

Conditions of strength of dilating non-cohesive soil

7.1 GENERAL PROPOSITIONS

The conducted research enables us to introduce adjustments to the conditions of strength during shear and triaxial compression. The main conclusion is that dilatancy, when constrained, constitutes a factor of stress condition and strength. The angle of internal or contact friction is not a constant quality for the given soil, but depends on the conditions of dilating. Allowing for these conditions opens up possibilities for finding correlative dependencies between strength and deformative parameters of granular medium. Below the readers will find an analysis of conditions of limiting equilibrium and strength factors of dilating non-cohesive soils during shear and triaxial compression.

7.2 ULTIMATE RESISTANCE TO SHEAR

According to Sections 4.2 and 5.2 conditions of strength for dilating soil during shear takes the following form:

$$\tau_u \leq \tau_d + \sigma_{n_o} \operatorname{tg} \varphi'_{(c)} \tag{7.1}$$

If we substitute (4.7), (4.8) and (4.10), (4.11), we can write down empirical expressions respectively for axis-symmetric and flat sums (shear of round and flat surface):

$$\tau_u \leq a\sqrt[3]{\frac{E}{(1+\nu)r}} + \sigma_{n_o} \operatorname{tg}\left[\varphi_\mu + (\varphi_o - \varphi_\mu)\exp-\alpha\frac{E}{(1+\nu)r}\right] \tag{7.2}$$

and

$$\tau_u \leq a\sqrt[3]{\frac{E}{(1-\nu^2)\omega b}} + \sigma_{n_o} \operatorname{tg}\left[\varphi_\mu + (\varphi_o - \varphi_\mu)\exp-\alpha\frac{E}{(1-\nu^2)\omega b}\right] \tag{7.3}$$

These expressions reveal the physical sense of interrelationship of strength and deformative characteristics of soil. What we receive is combining in one equation strength parameter $\varphi'_{(c)}$, characteristic of initial stress condition σ_{n_o}, deformative characteristics E, ν, and geometrical qualities r, b, ω which reflect the shear surface parameters. Additionally, the quality $\varphi'_{(c)}$ via the angle of intergranular friction φ_μ indirectly takes into account the type of mineral of soil grains, and via the gear angle $\varphi_g = (\varphi_o - \varphi_\mu)$ – the influence of density of packing, size, roughness and homogeneity of grains.

The exclusion of dilatancy reduces expression (7.2) and (7.3) to the usual Coulomb equation, i.e.

$$\lim_{\substack{E \to 0 \\ r,b \to \infty}} \tau_d = 0, \tag{7.4}$$

$$\lim_{\substack{E \to 0 \\ r,b \to \infty}} \varphi'_{(c)} = \varphi_o, \tag{7.5}$$

consequently,

$$\tau_u = \sigma_{n_o} \operatorname{tg} \varphi_o. \tag{7.6}$$

Let it be noted, though, that by expression (7.1) we have only a new understanding of Coulomb's law (1.3), according to which normal pressure and the angle of internal friction are variable qualities which depend on the conditions of realization of critical density.

We shall now analyze the influence of constrained dilatancy on resistance to shear.

In empirical equations (7.2) and (7.3) there are two groups of variables which determine the realization of dilatancy, deformative and geometrical, namely, on the one hand – modulus of elasticity E, Poison coefficient ν; and on the other hand – dimensions of shear surface (radius r, width b and coefficient ω). It is clear that, when the given functions are employed, the correct calculation of these qualities determines the reliability of the calculated resistance to shear.

In the case of the sums for contact friction, parameters r, b, ω are known in advance, and the problem of modelling shear conditions is reduced to determining E and ν. On the contrary, when stability is calculated, or in the case of 'soil-against-soil' shear the dimensions of shear zones and their orientation demand special measuring and determining, and the task becomes much more complicated. As for the condition of constrained dilatancy, it can emerge when shear zones appear in a landslide slope, when the sliding surface is not formed and the dilatant thrust unwedging it appears only in one separate section. Apparently, there is a necessity to make a detailed study of the influence exerted by dilatancy in such zones on the very process of formation of lines of sliding. We shall turn, though, to a simpler case of contact shear and analyze the influence of deformative and geometrical characteristics on strength parameters and resistance to shear.

98 *Conditions of strength of dilating non-cohesive soil*

7.2.1 *Influence of deformative characteristics of massif*

Let us assume, according to Section 2.5, that the value of modulus of elasticity approximately corresponds to eight values of modulus of deformation. Let us turn to Table A1.1 SNIP 2.02.01-83 'Foundations of buildings and constructions', according to which the recommended values of modulus of deformation constitute from 35 to 50 MPa for dense sands and from 18 to 40 MPa for sands of medium density, with mean values being respectively 44.5 and 29.0 MPa. We shall assume for further analysis, proceeding from (2.21), that the values of models of elasticity $E_1 = 44.5 \times 8 = 356$ MPa and $E_2 = 29.0 \times 8 = 232$ MPa.

Table A1.6 represents calculated values of coefficients of elastic resistance K according to formulae (2.2) and (2.3) for the given mean values of the modulus. Coefficient ω was taken from Table A1.2. Poison coefficient was assumed to be equal to 0.3. The values of τ_d and $\varphi'_{(c)}$ for control-mix sands (dense and medium-dense), were calculated taking into account the data from Table A1.6 according to (4.6) and (4.9) with coefficients a and α from Table 5.4, are given in Tables A1.7-A1.12.

A comparison of the values of τ_d and φ'_c (with both values of the modulus) shows how the precision, with which the modulus is measured, can influence the possible error in calculating strength parameters (see Table 7.1). As it turns out, when the values of E_1 and E_2 differ by 124 MPa, or 35%, the difference in the calculated values of τ_d does not exceed 21%, and tg φ'_c –7%. This fact prompts us to draw the conclusion that possible errors in estimation the modulus of elasticity caused by imperfect measurement techniques are not really an obstacle to employing this quantity in taking into account the influence of dilatancy on soil strength.

7.2.2 *Influence of dimensions and form of the shear surface*

An analysis of Table A1.7-A1.12 and of function $\tau_d, \varphi'_{(c)} = f(r, b)$ in Fig. 7.1 indicates that a major influence upon calculated strength parameters is enjoyed by the shear surface dimensions, which determine the values of coefficient of elastic resistance with the given modulus of elasticity. A drastic change of τ_d and φ'_c is observed within the range of values of r and b from 0.1 to 1.5 m. The corresponding values of τ_d for coarse sand differ by 2.9 times, φ'_c – by 1.2-1.4 times. With a

Table 7.1. Maximum difference of values of τ_d and tg φ'_c within the range of values of modulus $E = 232$-356 MPa for control-mix sands (according to the data of Table A1.7-A1.12).

Parameter	Coarse sand		Medium sand		Fine sand	
	dense	medium-dense	dense	medium-dense	dense	medium-dense
τ_d	15 %	14 %	15 %	20 %	16 %	21 %
tgφ'_c	7 %	7 %	7 %	4 %	4 %	4 %

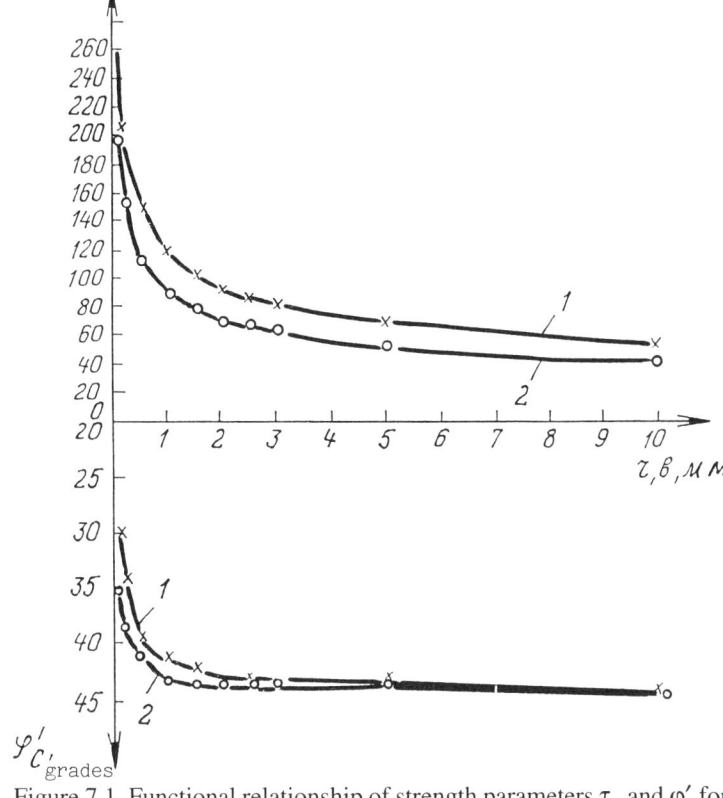

Figure 7.1. Functional relationship of strength parameters τ_d and φ'_c for coarse dense sand to dimensions and form of shear surface. 1. for cylindrical surface; 2. for flat surface with $L/b = 100$.

further increase of shear surface size the influence of constraint of dilatancy gradually decreases. For the case of an axis-symmetrical sum it can also be inferred that dilatancy constraint is in direct proportion to the curvature of the shear surface $i = 1/r$. In connection with this, it appears possible to 'regulate' the obtained values of resistance to shear in the process of designing deep foundations, for example, while recommending this or that pile diameter (within certain boundaries, it will be understood). For instance, for a pile with diameter 0.2 m the value of τ_u, in coarse sand of medium density with $E = 232$ MPa and mean pressure on the skin surface equalling 100 kPa, will constitute, according to (7.1), 174 KPa. The same quantity for piles with a diameter of 1 m will turn out to be equal to 143 KPa, which is 22% less. On the other hand, when resistance to shear is calculated by Coulomb expression of the equation (7.6) without allowing for the factor of dilatancy, it will amount to only 87 KPa, the error for the pile with a diameter of 0.2 m being 100%, for the pile with a diameter of 1.0 m – 64%. Experimental data proving the influence of pile diameter on the mobilized value of τ_u are represented in Chapter 8.

100 Conditions of strength of dilating non-cohesive soil

Tables A1.7-A1.12 also allow us to trace the influence of the shear surface form. The increase of the ratio of the length of shear surface L to its width b leads to a decrease of the dilatant component τ_d and an increase of the angle of contact friction φ'_c. Fig. 7.2 represents charts illustrating these functions for different values of the shear surface width/breadth. Values calculated for the axis-symmetrical sum are plotted for the sake of comparison.

With the ratio L/b equalling 2-3 the values of τ_d for cylindrical shear surface begin to considerably exceed those calculated for the flat surface, while the angles φ'_c become smaller. Let it be remembered that an increase of the ratio L/b, i.e. of the shear area and, correspondingly, of coefficient ω, signifies a simultaneous decrease of coefficient of elastic resistance, i.e. of resistance to dilatancy. According to Table A1.7-A1.12 the values of τ_d for cylindrical shear surface with $r = b$ are higher by 11% if $L/b = 10$, by 21% if $L/b = 50$, and by 24% if $L/b = 100$.

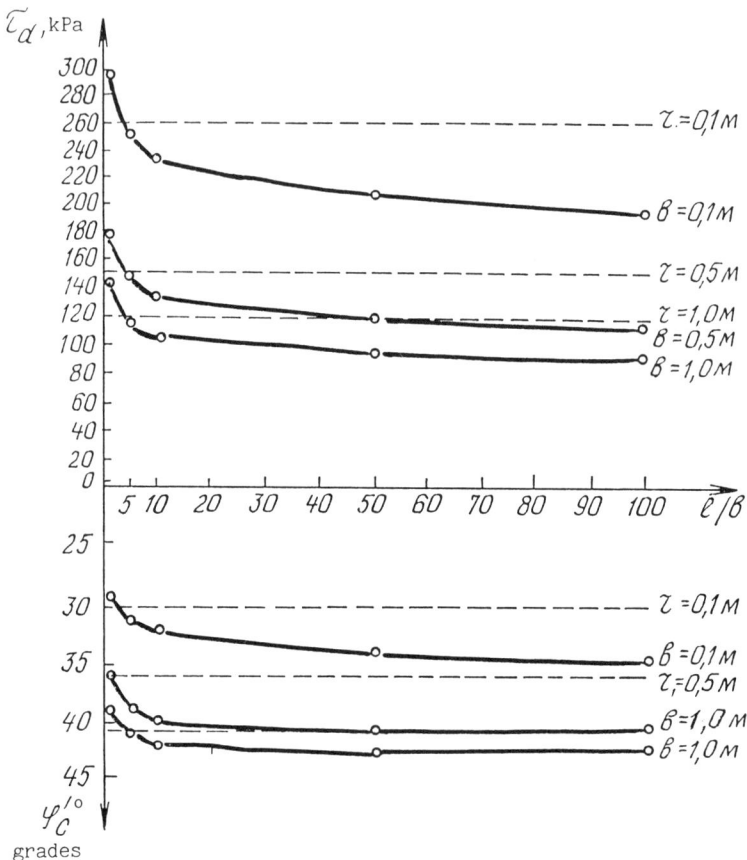

Figure 7.2. Functional relationship of strength parameters τ_d and φ'_c for coarse dense sand to the ratio of sides of a flat shear surface in comparison with the values for a cylindrical surface (featured by dashes).

7.3 ULTIMATE STATE DURING TRIAXIAL COMPRESSION

According to (6.3) the soil strength condition during triaxial compression can be expressed by the equation

$$\frac{\sigma_{(0)1} - \sigma_{(0)3}}{\sigma_{(0)1} + \sigma_{(0)3} + 2\sigma_d} = \frac{\sigma_{(0)1} - \sigma_{(0)3}}{\sigma_{(0)1} + \sigma_{(0)3} + 2\tau_d/\operatorname{tg}\varphi'} \leq \sin\varphi' \qquad (7.7)$$

From this condition the principal stresses in maximum and minimum ultimate states will be expressed, respectively, as

$$\sigma_1^{(max)} = \sigma_{03}\operatorname{tg}^2\left(45° + \frac{\varphi'}{2}\right) + 2\tau_d\operatorname{tg}\left(45° + \frac{\varphi'}{2}\right) \qquad (7.8)$$

and

$$\sigma_3^{(min)} = \sigma_{01}\operatorname{tg}^2\left(45° - \frac{\varphi'}{2}\right) - 2\tau_d\operatorname{tg}\left(45° - \frac{\varphi'}{2}\right) \qquad (7.9)$$

where σ_{03} and σ_{01} are respectively initial (before loading) principal normal stresses in massif (see Section 6.1).

Substituting functions (4.7), (4.8) and (4.10), (4.11) we can write down empirical expressions:
– for axis-symmetrical sum

$$\sigma_1^{(max)} = \sigma_{03}\operatorname{tg}^2\left[45° + \frac{\varphi_\mu + (\varphi_0 - \varphi_\mu)\exp - \alpha\frac{E}{(1+\nu)r}}{2}\right]$$

$$+ 2a\sqrt[3]{\frac{E}{(1+\nu)r}}\operatorname{tg}\left[45° + \frac{\varphi_\mu + (\varphi_0 - \varphi_\mu)\exp - \alpha\frac{E}{(1+\nu)r}}{2}\right] \qquad (7.10)$$

and

$$\sigma_3^{(min)} = \sigma_{01}\operatorname{tg}^2\left[45° - \frac{\varphi_\mu + (\varphi_0 - \varphi_\mu)\exp - \alpha\frac{E}{(1+\nu)r}}{2}\right] - 2a\sqrt[3]{\frac{E}{(1+\nu)r}}$$

$$\operatorname{tg}\left[45° - \frac{\varphi_\mu + (\varphi_0 - \varphi_\mu)\exp - \alpha\frac{E}{(1+\nu)r}}{2}\right] \qquad (7.11)$$

– for flat sum for maximum ultimate state (bulge) or minimum ultimate state (unloading) at contact with a flat surface

$$\sigma_1^{(max)} = \sigma_{03}\operatorname{tg}^2\left[45° + \frac{\varphi_\mu + (\varphi_0 - \varphi_\mu)\exp - \frac{E}{(1-\nu^2)\omega b}}{2}\right]$$

$$+ 2a\sqrt[3]{\frac{E}{(1-\nu^2)\omega b}} \operatorname{tg}\left[45° + \frac{\varphi_\mu + (\varphi_0 - \varphi_\mu)\exp - \alpha\frac{E}{(1-\nu^2)\omega b}}{2}\right] \quad (7.12)$$

and

$$\sigma_3^{(\min)} = \sigma_{01} \operatorname{tg}^2\left[45° - \frac{\varphi_\mu + (\varphi_0 - \varphi_\mu)\exp - \alpha\frac{E}{(1-\nu^2)\omega b}}{2}\right.$$

$$\left. - 2a\sqrt[3]{\frac{E}{(1-\nu^2)\omega b}} \operatorname{tg}\left[45° - \frac{\varphi_\mu + (\varphi_0 - \varphi_\mu)\exp - \alpha\frac{E}{(1-\nu^2)\omega b}}{2}\right] \quad (7.13)$$

In these expressions, as in (7.2) and (7.3) the parameters of the soil medium are taken into account, introduced body or contact surface and stress condition, namely: density, size (a, α) and mineral composition of grains (φ_μ), their roughness (φ_g), deformative characteristics (E, ν), dimensions of the failure area (r, b, ω), initial principal stresses σ_{03}, σ_{01}.

The exclusion of constraint on dilatancy, i.e. the fulfilment of conditions (7.4) and (7.5), reduces expressions (7.10-7.13) to their traditional form for non-cohesive soils. The condition of strength (7.7) in this case returns to the Rankine equation, which (using the designations assumed in Section 6.1) is written down as

$$\frac{\sigma_{01} - \sigma_{03}}{\sigma_{01} + \sigma_{03}} \leq \sin\varphi_0 \quad (7.14)$$

Respectively, principal stresses in maximum and minimum ultimate states

$$\sigma_{01}^{(\max)} = \sigma_{03} \operatorname{tg}^2\left(45° + \frac{\varphi_0}{2}\right)\varphi_0 \quad (7.15)$$

and

$$\sigma_{03}^{(\min)} = \sigma_{01} \operatorname{tg}^2\left(45° - \frac{\varphi_0}{2}\right)\varphi_0 \quad (7.16)$$

We shall now analyze the major factors influencing soil resistance in maximum (passive) and minimum (active) ultimate states with the help of experimental data for control-mix sands. While interpreting the results of this analysis, the passive pressure of soil for axis-symmetrical sum and the resistance to internal bulge can be regarded, for example, as the maximum resistance to the expanding of a cylindrical hole, while active pressure can be treated as the pressure of soil along the walls of the hole on its casing (see Chapter 8). For a flat sum it can be classified as the resistance of soil under the lower end of a footing (trench foundation) and the pressure on the skin surface of a 'diaphragm wall' during cutting of soil from one of its sides (see Chapter 10). Conditions close to free dilatancy can exist for active pressure of embankment of the retaining wall, for passive pressure of bond of the wall at the moment of bulge. But even in this case dilatancy introduces its own adjustments, which transpire when theoretical calculations are compared with in-situ measurements. At the same time free dilatancy is impossible, for example, when a hole is expanded during a pressiometrical test or cement mortar is injected into the hole.

Tables A1.13-A1.18 represent calculated values of passive (σ^{max}) and active σ^{min}) pressures of control-mix sands (dense and medium-dense) (see Section 4.1) for axis-symmetrical sum. Tables A1.19 and A1.20 give the values of σ^{max} and σ^{min} for flat sum and coarse sand (dense and medium-dense). The calculation was done for the mean modules of elasticity of massif $E = 356$ and 232 MPa assumed in Section 7.2.1.

While considering the tables one notes the negative values of active pressure, which is connected with the fact that the dilatant component of the expression (7.9) $- 2\tau_d$ tg $(45° - \varphi'/2)$ turns out to be larger than the component of initial stress σ_{01} tg^2 $(45° - \varphi'/2)$.

Physically, the negative active pressures can be interpreted as a reserve of strength of non-cohesive soil from mutual unwedging of grains near the failure surface. For cylindrical holes these are the stresses which prevent their walls from crumbling down even in the case of absence of casing – the arch effect (see Chapter 8). We should stress, though, that in expressions (7.8) and (7.9) we took into account only the factor of dilatancy, but not the influence of capillary and filtration forces or humidity. A more detailed analysis of such influence requires special additional research.

Let us turn now to the major factors of maximum and minimum stress states of granular medium.

7.3.1 *Influence of deformative characteristics of massif*

Figs 7.3 and 7.4 represent on the example of coarse sand the dependence of passive and active pressures on coefficient of elastic resistance in comparison with the values calculated without allowing for the influence of dilatancy. A comparison of these charts yields a considerable difference between the obtained values of σ^{max} and σ^{min}.

From the chart of function $\sigma^{max} = f(\sigma_{03})$ it is clear that the values calculated by the formulae (7.8) and (7.15) coincide only once, when initial lateral pressure corresponds to the critical density of soil. For each value of K there is only one point of intersection of straight lines $\sigma^{max} = f(\sigma_{03})$. This point corresponds to critical initial pressure σ_{03}^{cr}, the sense of which is analogous to that given to critical initial normal pressure during shear $\sigma_{n_o}^{cr}$ (see Sections 4.2 and 4.8).

The critical density at which soil failure takes place can be more or less than initial density. In the chart (Fig. 7.5) we observe dilatancy to the left of abscissa

104 *Conditions of strength of dilating non-cohesive soil*

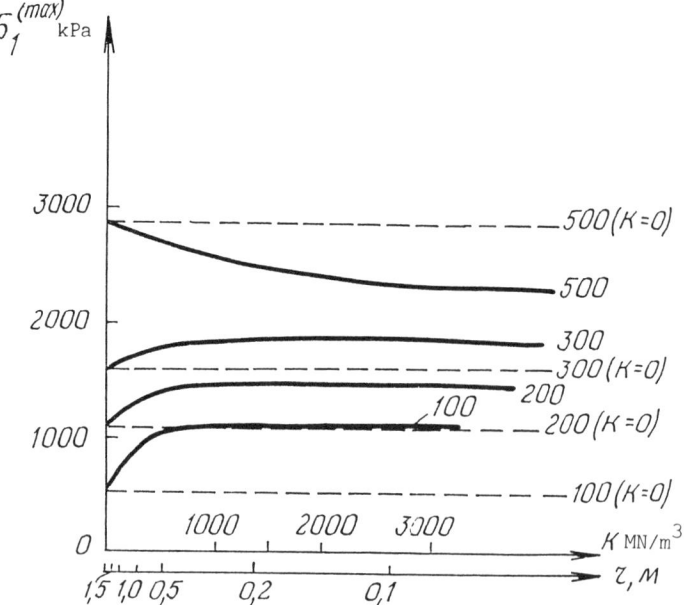

Figure 7.3. Functional relationship of values of $\sigma_1^{(max)}$ for coarse dense sand with $E = 356$ MPa to the values of coefficient $K = E/(1 + \nu)r$ and of radius r. Figures give the initial values of lateral pressure σ_{03}; dashes represent the values of σ_1 for the conditions of free dilatancy, i.e. $K = 0$.

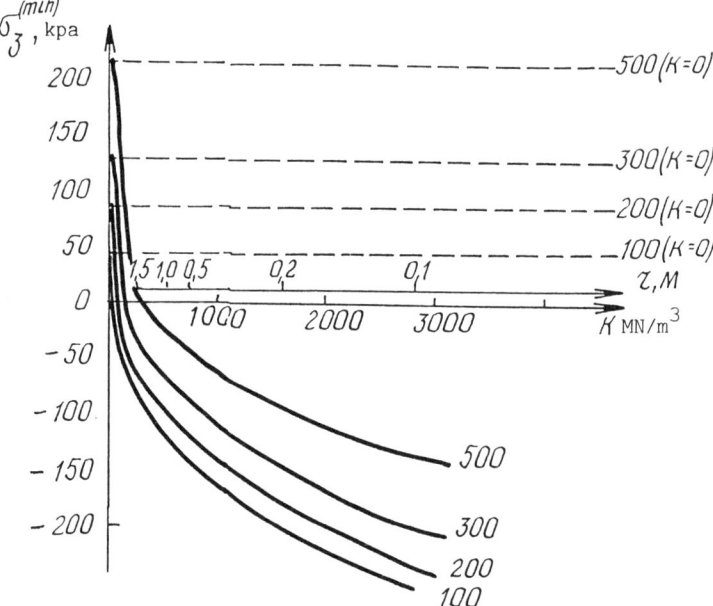

Figure 7.4. Functional relationship of values of $\sigma_3^{(min)}$ for coarse dense sand with $E = 356$ MPa to the values of coefficient of elastic resistance $K = E/(1 + \nu)r$ and of radius of the surface of the minimum ultimate state area r. Figures give the initial values of the principal normal stress σ_{03}; dashes represent the values of σ_3 for the condition of free dilatancy, i.e. $K = 0$.

σ_{03}^{cr}, and contraction to the right of it. In Fig. 7.3 the excess of σ_{03}^{cr} corresponds to the change of curvature of functions $\sigma^{max} = f(K)$ to its opposite. Hence, formula (7.15) in the case of constrained dilatancy gives underread values of passive pressure when $\sigma_{03} < \sigma_{03}^{cr}$, and overread values when $\sigma_{03} > \sigma_{03}^{cr}$.

Unlike the values of passive pressure, those of active pressure, with free and constrained dilatancy and $\sigma_{01} > 0$, do not coincide. An intersection of the straight lines $\sigma^{min} = f(\sigma_{01})$ is possible if they are continued into the area of negative initial pressures. This is devoid of physical sense due to the impossibility of resistance of non-cohesive medium to tension.

A comparison of the values of pressures σ^{max} within the range of E = 356-232 MPa (by the data from Tables A1.13-A1.18) shows inconsequential influence of the modulus values on the pressures (within the limits of 2-5%). A slightly larger difference is observed for active pressure, but even in this case the difference of the values of σ^{min} fits within the 25% limit. Thus, in the case of triaxial compression the errors of methods of calculating the soil's modulus of elasticity do not constitute a hindrance for a successful application of the model put forth in Section 2.4.

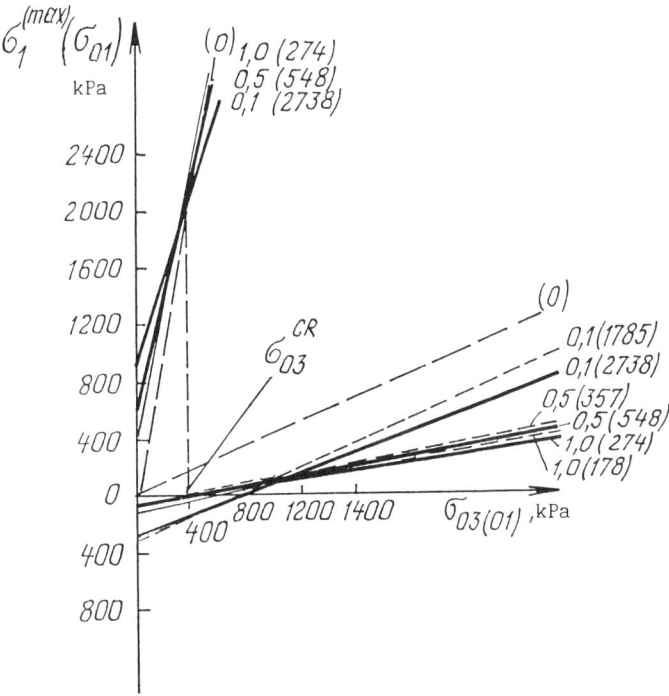

Figure 7.5. Functions $\sigma_1^{(max)} = f(\sigma_{03})$ and $\sigma_3^{(min)} = f(\sigma_{01})$ (respectively the upper and the lower groups of straight lines) for coarse dense sand with E = 356 MPa and r = 0.1; 0.5; 1.0 m (values of K are given in brackets). Dashes represent the values of σ_3 with E = 232 MPa (for σ_1 the straight lines almost merge). Straight lines (0) reflect the values for the case of free dilatancy.

7.3.2 Influence of the failure area dimensions

An increase of the size of the failure area lowers the influence of dilatancy on the obtained values of passive and active pressures. This is clearly seen in Fig. 7.5 and Tables A1.13-A1.18.

The charts in Figs 7.3 and 7.4, where the abscissa (axis K) is matched with the value of radius r, also allow to assess the influence of dilatancy constraint. It can be inferred that, other things being equal, the influence of dilatancy on the soil resistance in maximum and minimum ultimate states is higher, when the size of the failure area is smaller (i.e. when K is higher).

7.3.3 The grain strength factor

As obtained data testifies, during triaxial compression very high stresses develop in dilating non-cohesive soil. So, even when initial principal stress σ_{03} = 100-200 kPa, the passive pressure for coarse medium-dense sand with r = 0.2 m reaches 714-1053 kPa, for fine sand – 424-712 kPa. With higher values of σ_{03} these figures go up further.

At the same time, it is assumed that failure of quartz grains takes place at pressures exceeding 700 kPa. It is logical to suppose, then, that this failure will be more substantial when grains are larger and more homogenous.

When $\sigma_{03} > \sigma_{03}^{cr}$, failure is accompanied by contraction. In the case of dense soil, where all grain repacking capabilities are exhausted, contraction is possible only at the expense of breaking the large grains into finer ones. New values of strength parameters will correspond to the new grain size. Taking into account the high level of pressures which develop during triaxial compression of dilating soil, these phenomena can hardly be regarded as minor. They have to be researched specially in triaxial instruments which allow to reach working pressures exceeding 1.5-2.0 MPa and at the same time do not let the liquid to leak out of the camera.

Due to the technical complexity of this task, another way can be taken in the first approximation. For example, lowering adjustments can be introduced to the values of τ_d and φ' when σ^{max} > 700 kPa. Thus, an introduction of 30% adjustment for τ_d and a 4° (but not less than φ_μ) adjustment of φ' would result in the case of coarse sand when σ_{03} = 200 kPa, r = 0.5 m in a 16% decrease of the calculated value of σ^{max}.

7.4 CONCLUSION

The Coulomb-Mohr strength theory in its traditional understanding is true only for the conditions of free dilatancy and critical initial density of a granular medium. Thus, in order to describe strength conditions for the case of constrained dilatancy, two major adjustments have to be introduced:

a) Normal pressures in the process of failure are not constant, but depend on the reaction of the surrounding massif to the accompanying volume deformations;

b) The angle of internal (contact) friction depends on the degree of realization of the grain-gear angle, which in its turn is determined by conditions of development of dilatancy.

According to this understanding, the *dilatancy can be taken as a physical property which unites strength and deformability of granular medium.*

The revealed close correlation of strength and deformative parameters opens up new possibilities for a mathematical description of granular medium strength which takes into account the main factors determining it (density, grain size, humidity, deformative characteristics and the dimensions of the failure area).

A number of possible directions for further research could be singled out:

1. Studying the role of dilatancy in the character of grainy medium failure: localization of shears along sliding surfaces, formation of bulge zones etc.

2. Refinement of the obtained mathematical functions for describing the grainy medium behaviour in the conditions of constrained dilatancy.

3. Development of the concepts set forth in this work for dealing with practical geotechnical tasks.

2. Deep foundations in dilating soil

CHAPTER 8

Constrained dilatancy as a factor of load-holding capacity of deep foundations

8.1 GENERAL PROPOSITIONS

This chapter treats certain properties of technologies in their interconnection with the load-holding capacity of piles, footings and injection anchors. These constructions are distinguished by similarity of technological methods and approaches to the design of their load-holding capacity represented by a broad spectrum of methods ranging from purely theoretical to completely empirical.

The construction of a pile, footing or anchor includes drilling a hole or excavating a cutting in the soil massif. This is accompanied by local changes of the soil's original physical and mechanical properties. Consequently, one of the main problems at the technology-choice and calculation stage is a correct evaluation of the actual parameters of base in the contact zone.

The indeterminate nature of changes in the soil properties in the process of drilling or cutting, filling a hole or trench with concrete, the shrinkage of concrete and other similar factors are usually considered an objective obstacle to working out universal calculation methods. At the same time there is a backlog of contradictions of a different nature testifying to the inapplicability of the traditional soil strength conditions to the calculation of deep foundations. For instance, values of contact friction by far surpassing the values expected in accordance with Coulomb's law are measured while testing injection piles and anchors. Values of average contact friction τ_s equalling 250-400 kPa, measured with values of normal pressure by an order of magnitude lower, are not infrequent. If we proceed from the condition of strength analogous to (7.1),

$$\tau_s = \sigma_{n_{oi}} \operatorname{tg} \varphi'_{ci}, \qquad (8.1)$$

where $\sigma_{n_{oi}}$, φ'_{ci} are average normal pressure and angle of contact friction for the given layer respectively, then this ratio signifies that this parameter of strength by far surpasses 45° (see Section 1.7). So far no satisfying explanation has been provided for this contradiction. But the attention of researchers is ever more persistently turning to dilatancy as a possible key to the problem.

A distinctive feature of modern technologies of deep foundations is their active

Table 8.1. Classification of pile types under study.

Pile type	Concise description of technology of constructing
1.	Piles installed in holes (drilled with/without the protection of a casing) by ordinary filling in with concrete accompanied/unaccompanied by vibration
2.	Piles installed in bore holes (drilled with the protection of bentonite suspension) by filling in with concrete using the tremie pipe method accompanied with vibration
3.	Piles constructed by 'hollow-stem auger' method
4.	Piles constructed using technologies 1-2 with consequent pressing of the base under the lower end by injection of cement mortar
5.	Piles constructed using technologies 1-2 with consequent grouting under the lower end and along the skin surface
6.	Piles installed in bore holes with the diameter of up to 200 mm by injection of cement mortar (injection- or 'micropiles')
7.	Piles constructed with displacement of soil (driven piles, filled piles constructed by driving of inventory pipe, vibro-stamping piles)

Table 8.2. Classification of trench foundations (footings) under study.

Foundation type	Concise description of technology of constructing
1.	Footings constructed by the diaphragm wall method under bentonite suspension, and filling the trench with concrete by the tremie pipe method
2.	Footings constructed using technology 1 with consequent compressing of soil under the lower end by grouting
3.	Footings constructed using technology 1 with consequent compression of soil by grouting both under the lower end and along the skin surface

effect on soil properties. Grouting as a means of compacting, stiffening and compressing is being employed more and more often and increasingly successfully. The arsenal of modern foundation engineering allows us not only to adequately use the natural soil properties, but also to modify them within a certain range. However, this demands full clarity in the actual mechanisms of soil behaviour and in determining the primary and secondary factors of bearing capacity.

Chapters 9-11 contain suggestions as to the methods of calculation of bearing capacity of piles, trench foundations and injection anchors on the basis of corrected conditions of strength by Coulomb-Mohr. To facilitate the task, these constructions are classified by the technology of their erection. Table 8.1 features the basic types of piles, whereas Table 8.2 contains information on footings. Let us further consider the main peculiarities of each type of technology, taking into account their influence on the properties of dilating soil.

8.2 BORE PILES OF TYPE 1

Piles of this type are the most widely used in post-Soviet foundation engineering. Available drilling equipment allows to make holes with the diameter of up to 0.8-1.2 m, rarely with larger diameters. The most widely employed type of equipment is ram-cable installations, more rarely drilling machines. In the USA piles with diameters of up to 3-5 m have long been used to transfer extensive loads (the Chicago caisson method). In Europe, on the contrary, the tendency is to make more frequent use of piles with diameters of up to 1.5-2 m.

The operation of extracting the borehole casing tends to become more complicated with the increase of the diameter due to considerable contact friction along the walls of the pipe. As a result, long piles and large-diameter piles are often constructed by merely filling in the casing which is not used further. This renders the technology excessively expensive and not competitive.

When soil is excavated according to the technology of Type 1, particularly with the use of ram-cable installations, there is a considerable loosening of soil along the hole walls and at the bottom (Fig. 8.1). Moreover, excavation is usually performed by a method when the bottom is excavated to the depth of 0.5-1.5 m with the consequent submerging (usually by ramming) of the borehole casing, which may lead to the formation outside of the casing not only of loosened soil, but empty spaces (caverns).

During the excavation soil crumbles and forms a loosened layer at the hole

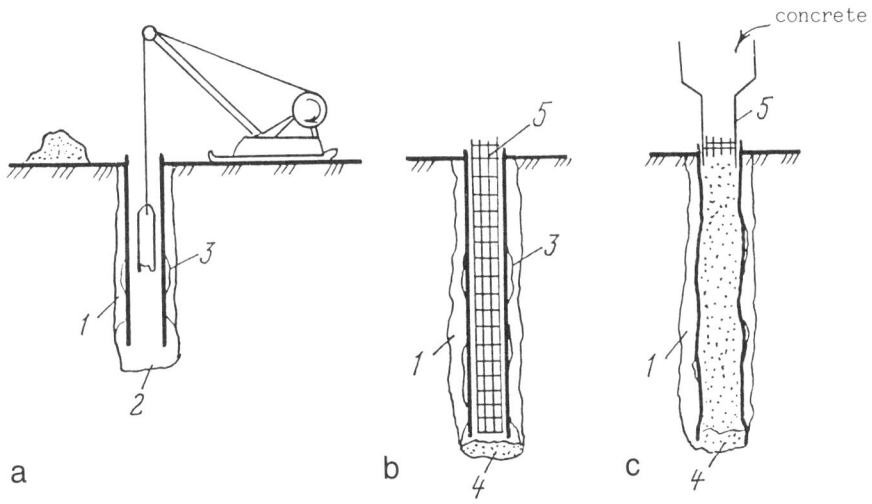

Figure 8.1. Sequence for constructing bore piles according to the traditional technology: a) excavation of a bore hole with the lagging behind of the casing; b) insertion of a reinforcing cage; c) concreting. 1. loosened soil along the hole's walls; 2. bottom of the hole; 3. caverns or empty spaces; 4. loosened soil or slurry accumulation at the bottom of the hole; 5. tremie pipe; 6. body of the pile.

114 *Constrained dilatancy as a factor of load-holding capacity*

Table 8.3. Calculated resistances along the skin surface of bore piles and cased piles τ_{si} according to SNIP 2.02.03-85.

Average depth of layer in the soil, m	Values of τ_{si}, kPa for sandy soils of average density		
	Coarse and medium	Fine	Silty
<5	26–39	16–28	11–20
5–10	39–46	28–32	20–24
10–15	46–50	32–36	24–27
15–20	50–55	36–39	27–29
20–25	55–60	39–43	29–31
25–35	60–70	43–49	31–35

bottom which can be several dozen centimetres thick. The cleaning out of this soil before the placement of concrete is fairly complicated and rarely, if ever, performed adequately. These drawbacks of technology of the first type cannot be compensated for by merely filling the hole in with concrete, because its weight and lateral pressure are not sufficient for any essential compaction of the loosened soil both at the bottom and alongside the walls of the hole. The situation is further aggravated by the shrinkage of concrete (Section 8.8). Consequently, the shifting of the loosened soil alongside the hole's lateral surface may be accompanied by contraction with simultaneous dropping of the lateral pressure. The result of this is low values of contact friction and the need for a considerable settling required to mobilise the base resistance. This settlement can exceed the limit acceptable for the given structure.

These facts are corroborated by low values of calculated resistance along the skin surface and lowering coefficients for bore piles in Tables 2 and 5 of Construction Norms and Rules of USSR (SNIP) (Sobolevsky, 1990). Calculated resistance for bore piles with working conditions coefficient $\gamma_{cf} = 0.7$ are given further in Table 8.3.

Technological inadequacies and great scatter of received values of contact friction and bearing capacity (very often measured for the same type of soil) also make it difficult to generalise empirical data. As for Table 2 in the abovementioned SNIP, it is compiled by processing data from numerous in situ experiments with driven piles conducted in various regions of the USSR. Consequently, it is strictly approximate and, as a rule, quotes underestimated values of the forces of soil contact friction. At the same time we presume that the recommendation to use Table 2 of SNIP or other norms of the same class for the design of injection piles is a priori erroneous.

It follows from SNIP that contact friction in sands from fine to coarse at depths ranging from 5 to 15 meters constitutes from 28-36 to 39-50 kPa, which fairly well agrees with the data given by Stocker (1986) (we borrowed these latter for Table 8.5). These values are close to those calculated according to Coulomb for the case of free dilatancy. Evidently, the loosening of the soil, unloading of the massif and the shrinkage of concrete jointly create conditions when only the soil contact layer participates in the shear, and dilatancy does not reach the point where it could provoke an elastic reaction in the surrounding massif.

8.3 BORE PILES OF TYPE 2 AND FOOTINGS OF TYPE 1

Technologies of this type allow to construct piles of great bearing capacity with diameters of 1.5-2.0 and more metres, and also deep footings of different configuration in plan. The sequence of constructing piles of Type 2 and footings of Type 1 is represented in Fig.8.2. After the installation of a conductor, the excavation is done by cable or hydraulic grab drilling auger or the flush-drilling method. Upon reaching the required depth the drilling instrument is extracted and the placement of cement is done by the tremie pipe. As far as no support is used with borehole casings, this technology allows to construct both round and multi-plane footings in Fig. 8.2.

The stability of the walls of the cutting in the process of excavation is ensured by the hydrostatic pressure of the weight of bentonite suspension and the hydrodynamic pressure of the incoming flux of filtrating water (see Section 8.8). Meanwhile a so-called filtrated 'cake' is formed in the walls of the cutting. The particles of suspension permeate the pores of the soil where due to ticsothropic properties they are transformed into 'jelly' and are dehydrated.

The character, thickness and resistance to shear of the clay contact layer are determined by the following factors (Stocker & Bauer, 1979):
 – grain size and soil permeability;
 – type of bentonite (fineness of milling);

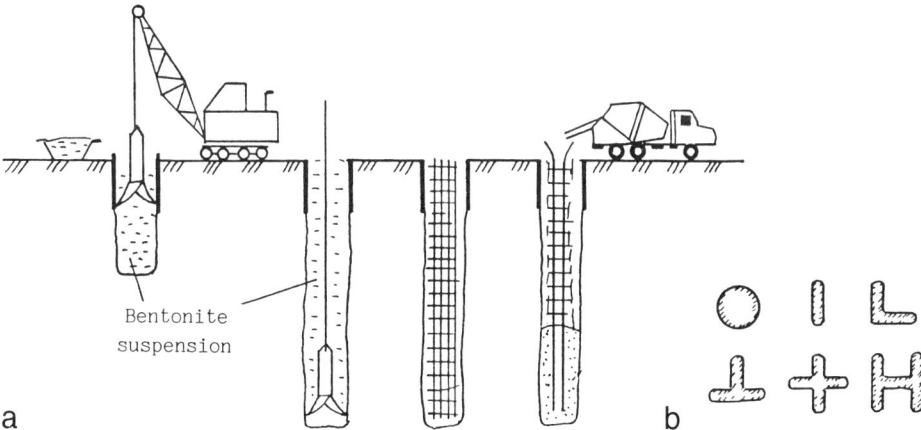

Figure 8.2. Sequence for constructing bore piles and footings with soil excavation under the protection of bentonite suspension (a), and the possible configuration of the constructions in plan (b).

Figure 8.3. Clayey contact layer near the wall of a hole (Stocker & Bauer, 1979). 1. pile concrete; 2. filtrated cake; 3. sandy soil.

- concentration of bentonite in the suspension;
- degree of suspension contamination;
- differential of hydrostatic pressure of the suspension and ground water;
- time during which the suspension remains in the hole.

To complete this list of factors we must add the influence of the incoming filtrating flux.

The thickness of the clayey contact layer in sand is usually 5-20 mm, in coarse gravel – 10-30 mm. A photograph of such a layer near the contact surface of a trench wall is represented in Fig. 8.3.

Vagueness in the evaluation of properties of the clay layer generates apprehensions as to the potential decrease in its strength as compared to natural soil. A special research to this effect was conducted by Farmer et al. (1971). According to this research, shear strength of bentonite suspension, fresh concrete and the filtrated layer are in the ratio of 1:15:120, i.e. fresh concrete is capable of displacing the suspension, but not the filtrated 'cake'. Cernak (1973) established that the properties of the clayey layer can range, depending on the distance from the filtrating surface, from the properties of plastic clay to those of suspension. The resistance of this layer to shear increases with the growth of the pressure of suspension. The data of Černak's laboratory experiments are represented in Fig. 8.4. As the author established, friction along lateral surface between sand and concrete does not change under the influence exerted by the suspension upon the soil for a period of time of up to 18 hours. At the same time, such influence exerted for 168 hours caused a drop in the angle of contact friction from 33 to 14 grades. Therefore, the decisive factor is the length of time during which the soil is exposed to the influence of the suspension, and this time should be reduced to minimum.

Similar results were obtained by Farmer et al. (1971). As a positive feature of this technology they note extra roughness of the hole's walls achieved as a result of the grab cutting, and also the fact that, due to the thrust influence of the suspension, soil is not subdued to uploading. Geffen & Amir (1971) also consider a lesser unloading of the soil during the drilling to be the cause of higher bearing capacity of piles of Type 2 as compared to piles of Type 1.

Excavation under the bentonite suspension allows for reduction of the accumulation of loosened soil at the bottom of the hole in comparison with drilling under borehole casing. The determining factors here are the time during which the suspension remains in the hole, its density and degree of contamination. In any case, similarly to the technology of Type 1 piles and footings, a special cleaning of the bottom before concreting remains highly preferable.

The experience of construction of piles of Type 2 is summarised in the Norms of Great Britain (1975) which contain the following requirements for the technology:

– In order to ensure perfect quality of a pile, its minimal diameter should be not less than 600 mm;
– Water-cement ratio of the concrete mixture should not exceed 0.6;
– In the progress of the work the quality of the suspension (density, viscosity,

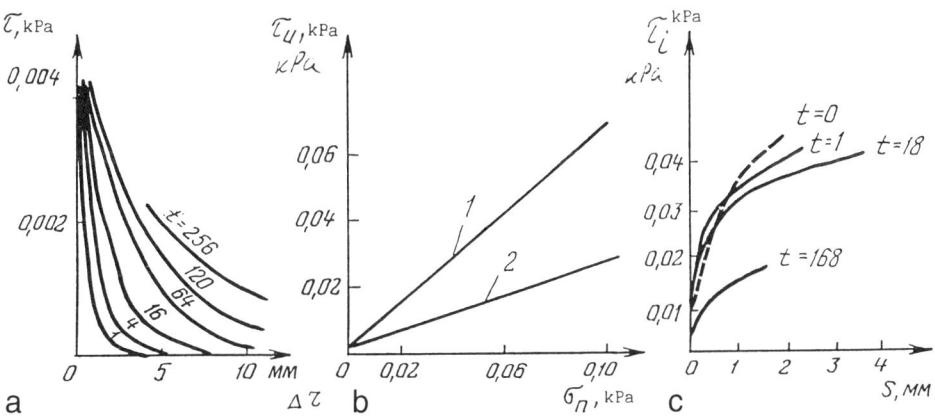

Figure 8.4. Data from laboratory experiments conducted by Černak (1973). Functions: a) 'resistance to shear τ – distance from the hole's wall Δr' (suspension in the hole for 1,4 to 256 hours); b) 'maximum resistance to shear τ_u – normal presure σ_n': 1. soil; 2. clayer filtered layer; c) 'friction along lateral surface τ_i – shear strain S_i' (suspension in the hole for 0,1 to 168 hours).

Table 8.4. Values of contact friction for piles and footings constructed using technology 2 (Stocker, Bauer, 1979).

Authors	Soil	Section of pile or trench foundation, mm	Contact friction τ_s (kPa)	Settlement or shear s, mm	Notes
Černak (1973)	Sand	150 × 440	40	4	Laboratory shear test with lateral pressure equalling 60 kPa; soil under suspension for 18 hours
Farmer et al. (1971)	Sand	Ø 150	120	5	Model test with lateral pressure equalling 70 kPa; soil under suspension for 5 hours
Geffen & Amir (1971)	Fine sand	Ø 650	70-120 100-130	5 10	Values calculated on the basis of the chart 'load-settlement' of the pile
Fernandez-Renau (1965)	Sand from medium to gravely	Ø 630	52 58	10 20	Pulling-out test, mean value along the pile length
Van Soos (1972)	Gravel, fine sand	2200 × 600	13-75	20	Pulling-out test, soil under suspension for various times
Reese et al. (1973)	Sand	Ø 760-920	100-175	10	Mean values for 3 sites
Bauer (1974)	Sand	Ø 1300	110-130	10	Transfer of force only in sand layers
Keinberger (1975)	Gravel	1500 × 500	74	10	Mean value along the pile length

Table 8.5. Values of contact friction for piles of type 1, 3, 5 (according to Stocker, 1986).

Pile type	Minimal and maximal values of τ_s, kPa with settlement			
	10 mm		30 mm	
	Sand	Gravel	Sand	Gravel
1	min 40-70 max 90-180	70-105 180-220	70-120 200-255	120-165 255-305
3	min 75-135 max 175-235	135-175 235-285	105-150 215-275	150-200 275-325
5	min 80-145 max 220-280	145-185 280-335	140-215 315-380	215-265 -

resistance to shear and the value of pH) should be under periodic surveillance;

– While using regenerated suspension it is necessary to control the sand content (samples should be taken directly at the bottom of the hole).

The question concerning the quality of suspension and the construction technology have received adequate treatment in the Construction Norms of the Republic of Belarus (RSN 20-87). In the GDR Norms for diaphragm walls, special attention was given to the necessity to take into account the time during which the suspension influences the values of calculated resistance of soil strength parameters. According to these documents, the angle of contact friction must be diminished if the time during which the suspension remains in the cutting exceeds four hours. As for the negative influence of sedimentation on the resistance under the lower end of the pile, it can be avoided provided the suspension is of high quality, the content of sand particles in it is not higher than 3%, and the suspension remains in the cutting for less than two hours. A special cleaning of the base is also recommended.

The values of contact friction of the skin surface of piles of Type 2 as established by different authors are represented in Table 8.4.

The analysis of literature shows that there is no unified opinion as regards the influence of the bentonite suspension on the values of contact friction. As for the resistance under the lower end of the pile, it primarily depends on the cleanness of the bottom. The settlement required to develop the maximum resistance is in proportion to the thickness of the loosened layer and accumulation of slurry under the lower end.

Thus, the quality of excavation under the protection of bentonite suspension is very sensitive to the strict adherence to the technological requirements. At the same time, a decrease of soil loosening incurred in the process of drilling constitutes a positive factor in comparison with the technology of constructing piles of Type 1. The values of contact friction for the shaft of piles of Type 2 are intermediary between analogous values measured for piles of Types 1 and 3 (see Tables 8.3-8.5).

8.4 BORE PILES OF TYPE 3

Piles constructed by the method of concreting through a hollow-stem auger are also sometimes termed screw piles, spiral piles, piles built with partial displacement of soil etc. At first such piles were constructed by the German firm 'Bauer Schrobenhausen' in 1974. At present piles of Type 3 are constructed with the length of up to 22 m and diameters of 400-1,000 mm (Bauer, Soletanche, 1990).

The sequence of constructing an auger pile is represented in Fig. 8.5. There are two varieties of this method. According to the first one (Fig 8.5a-e) the excavation of a hole is done while the drilling auger is screwed into the soil until the specified depth is achieved. A relatively small amount of soil is displaced in the process of

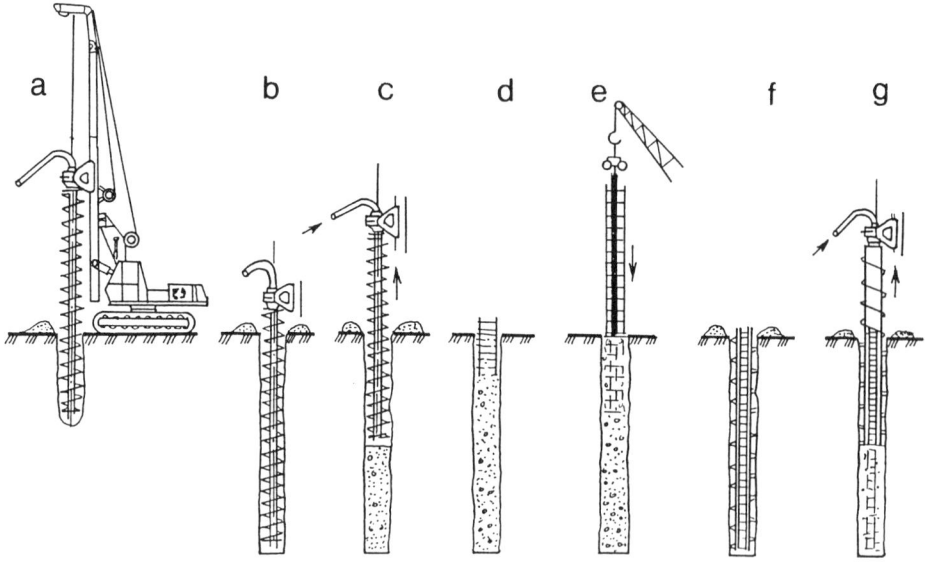

Figure 8.5. Sequence for constructing bore piles by the hollow-stem auger method.

drilling, this amount depends on the diameter of the auger's central core. Usually this diameter is 125 mm. The soil is partially displaced aside, being pressed into the hole's walls. Upon submerging of the auger to the specified depth, concrete mortar with setting of conus up to 50-60 cm is pumped under pressure through the auger's central core. The extraction of the drilling organ is effected simultaneously (Fig. 8.5c). To ensure a qualitative filling in of the hole, the pressure of the concrete must be equal to or exceeding the static pressure of the auger's weight (Stocker, 1986).

If total reinforcement is not imperative, a short cage is imbedded into the fresh concrete (Fig.8.5d). When total reinforcement is necessary, the cage is imbedded with the help of a vibrator attached to a stiff ruler.

The sequence of constructing the other variety of piles of Type 3 is represented in Fig. 8.5a, b, f, g. The difference lies in the design of the auger which has the exterior diameter of 500-800 mm and the core with the diameter of 400 mm and more. The soil excavation is performed by a drilling head at the end of the auger; the head is extracted through the core upon reaching the desired hole depth. Reinforcement is then introduced into the core, after which concrete mortar is fed in under pressure. The grout fills in the hole while the drilling organ is being extracted. The augers for both varieties of technology of Type 3 and the drilling mechanism mounted on a 'Kelly' bar are represented in Fig. 8.6.

As the experience of constructing such piles demonstrates, technology of Type 3 ensures a quality completion of pile shaft (Fig. 8.7), tight adherence of the concrete to the soil of the hole's contact surface, high production rate (up to

Figure 8.6. Designs of hollow-stem augers for soil-drilling (Stocker, 1986): a) by the method of screwing into the soil; b) by means of drilling head; c) drilling machine BG-26 equipped with an auger of the 'a' design (Stocker, 1986).

200-250 m of total pile length per shift). The completion of the pile shaft is effected simultaneously with the extraction of the drilling organ while the concrete mortar is piped under pressure. The latter process is accompanied by an intensive 'wringing' of free water into the soil, i.e. by compression and lowering of the mortar consistency. The difference in the volume of the hole and the fed concrete reaches from 10 to 30%.

Fig. 8.8a, b represents charts which allow for the comparison of the values of skin friction τ_s along pile shaft (piles constructed with the use of borehole casings (Type 1)) and with consequent injection of cement mortar along the lateral surface (Type 5). The charts show a range of values with the pile settlement equalling 10 and 30 mm respectively. These values were measured in situ both by taking separate measurements of pile, and by subtracting the former from the general value, tests were conducted in homogeneous soils. The values of τ_s according to the charts in Fig 8.8a, b are also given in Table 8.5.

It follows from Table 8.5 that if a pile is constructed with the use of technology of Type 3, the contact friction exceeds contact friction measured for piles of Type 1 on the average by 42% for sands and 34% for gravel with the settlement equalling 10 mm, and by 17% and 12% respectively with the settlement equalling 30 mm. The high values of τ_s for piles constructed with the use of the auger technology can be accounted for by the little loosening of the soil in the process of drilling, and by the high pressure of concreting (Stocker, 1986).

Figure 8.7. Pile shaft constructed by the hollow-stem auger method ('Bauer Schrobenhausen', Germany).

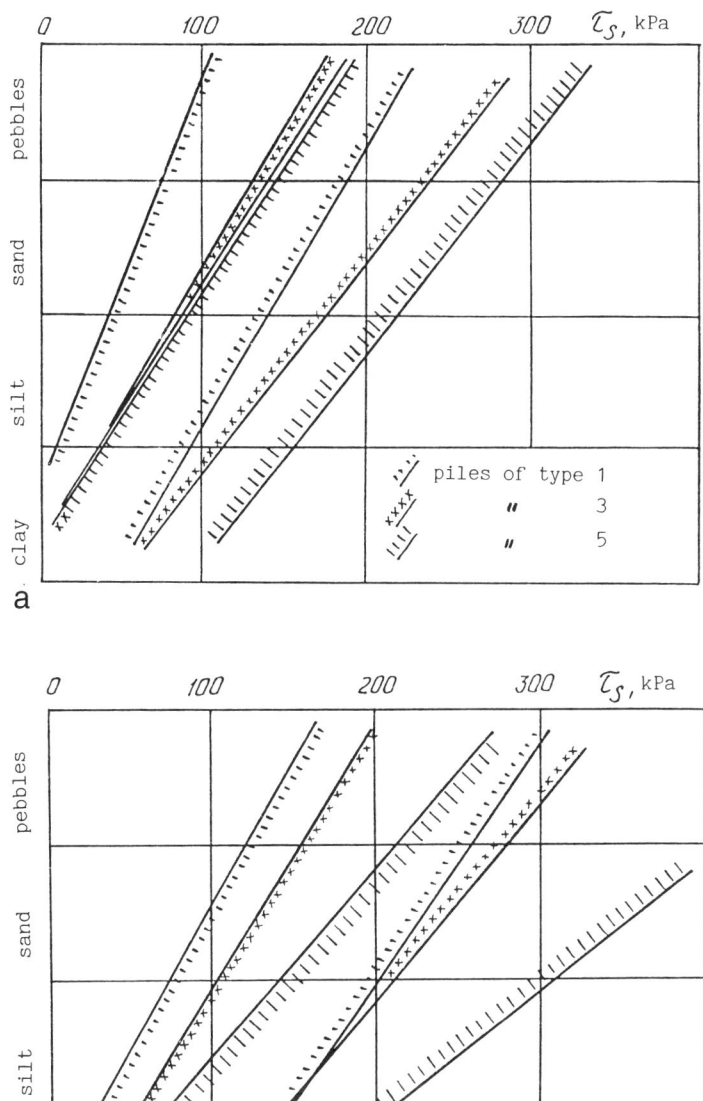

Figure 8.8. Mean values of lateral friction for piles constructed with the use of different technologies when the shear settlement equals: a) 10 mm; b) 30 mm (Stocker, 1986).

8.5 BORE PILES OF TYPES 4, 5 AND FOOTINGS OF TYPES 2, 3

The technology of piles construction with compression of the soil under the lower end and along the skin surface is based on the positive experience of grouting in constructing soil anchors. Both the technology of injection anchoring and the technology of compression of bore pile foundations were designed and used for the first time by the German firm 'Bauer Schrobenhausen'. As for the idea of employing injection for enhancing the bearing capacity at the lower end of the pile, it probably belongs to Grutman (1969). According to Stocker (1983), the special system for enhancing the bearing capacity of bore piles by means of injection was first tested in 1979. At present, piles of Types 4 and 5 are widely used in Europe and the Middle East (Lokau, Stoetzer, 1987). The use of this system has also gained wide acclaim for increasing the bearing capacity of footings (Davidson, 1988).

A method of increasing the bearing capacity used in Czechoslovakia, according to which a number of holes with the diameter of 80-120 mm are drilled under the lower end of a bore pile and then grout is injected into these holes, can also be regarded as a modification of pile technology of Type 4. In this method the bore pile is supported by a cluster of micropiles (Šimek & Verfel, 1989). Such technology, though, appears rather labour-intensive and expensive.

The sequence for constructing a pile using Type 5 technology is represented in Fig. 8.9. The distinction of Type 4 technology is the absence of injection of the

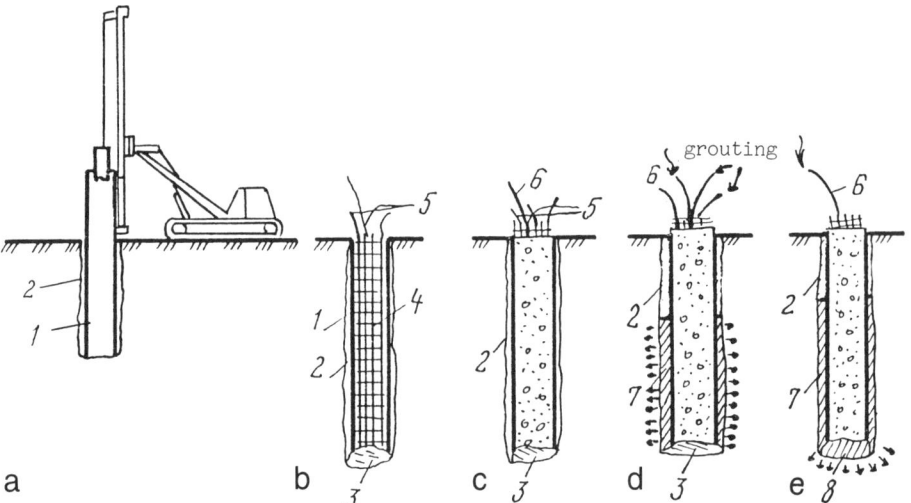

Figure 8.9. Sequence for constructing a pile base with compressing of soil at the lower end and along the lateral surface. 1. drilling organ; 2. hole's wall; 3. accumulation of slurry at the bottom of the hole; 4. reinforcement cage; 5. injection tubes of the skin surface; 6. injection tube of the pile's lower end; 7. pressed skin surface; 8. grouted base under the pile's lower end.

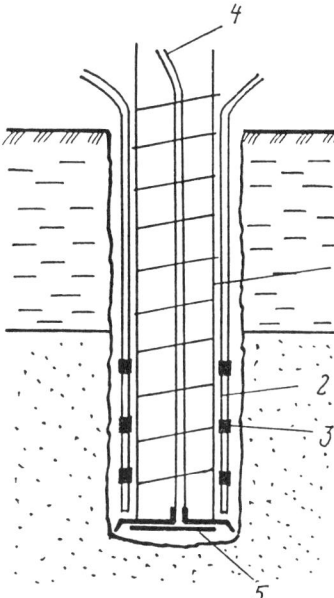

Figure 8.10. Schematic cage design for pile of type 5. 1. reinforcing rods; 2. injection tubes of the lateral surface; 3. valves; 4. injection tube of the lower end; 5. distribution plate with a skirt (Stocker, 1983).

skin surface. A cage with a system of injection tubes is lowered into a bore hole constructed by any method (Fig. 8.9b). Then the hole is filled in with concrete through a tremie pipe with or without vibration. After stiffening of the concrete 12-36 hours later an injection pump is joined up with the tubes of the external contour, and at water pressure in the order of 5 MPa the fresh concrete is disrupted along the pile contour opposite the valves (Fig. 8.9c). The rate of mortar flow constitutes usually 50-100 litres per valve. If necessary, injection can be effected several times. A rapid rise in pressure or refusal manifested by contact filtration of mortar on the surface around the external pile contour signifies termination of injection.

After injection along the hole contour is concluded and the grout has gained enough strength, injection under the lower end of the pile through a special injection tube is effected. The injection pressure P_{gr} in this case can reach 6 MPa and more (Stocker, 1983). The pile head rise, which should not exceed 1-2 mm, is monitored during injection. A higher rise can cause dropping of values of τ_S at loading. The rise of the pile is an indication of sufficient injection. Knowing the pressure of pile rise it is possible to get a rough idea of the pile's bearing capacity, i.e.

$$F = P_{gr} A, \tag{8.2}$$

where A is the area of the lower end of the pile.

According to (8.2) a pile with the diameter of 500 mm which began to rise at the pressure under the lower end $P_{gr} = 6$ MPa will have the approximate bearing capacity in the order of 1.18 MN.

126 *Constrained dilatancy as a factor of load-holding capacity*

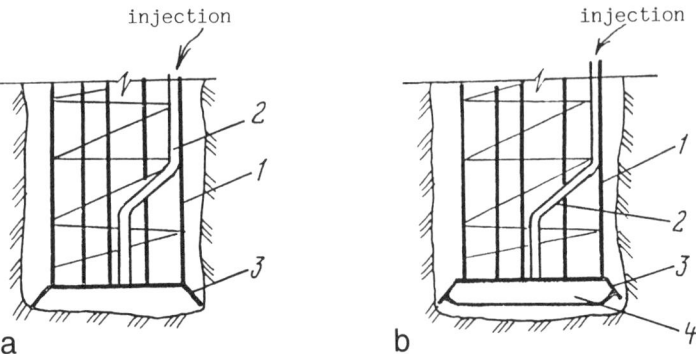

Figure 8.11. Design of bearing part for pile of type 5. a) with an open lower end; b) with a rubber membrane. 1. cage rods; 2. injection tube of the lower end; 3. distribution plate with a skirt; 4. rubber membrane.

Figure 8.12. Coating around the pile shaft after base compression by injection of cement grout: (a); pile shaft; (b) (Stocker, 1983; 'Bauer Schrobenhausen', Germany).

The reinforcement of pile of Type 5 is represented in Fig. 8.10. Injection tubes of the external contour are situated radially between the lengthwise reinforcement bars. The tubes' diameter is usually 20-25 mm. Along the tubes' length starting from the lower end of the reinforcement there are holes with a diameter of 6-10 mm with rubber valves or several layers of insulation tape. The distance between the holes is usually 0.5-0.8 m. In the upper part of the pile, not less that 3-4 m long, the holes in the tube are absent to ensure plugging.

The design of the bearing part of the pile is represented in Fig. 8.11. It can be completed as a steel sheet spacer with a skirt cutting into the soil, and be fitted with a rubber membrane. In the latter case the direct contact of the fed cement mortar with the soil and possible hydroruptures are excluded and the area to which the pressure is being distributed is known exactly.

The objective of the contour injection and of grouting under the lower end of the pile is compacting the soil. Injection not only eliminates the negative influence of soil loosening during excavation, but ensures a considerable compacting of the

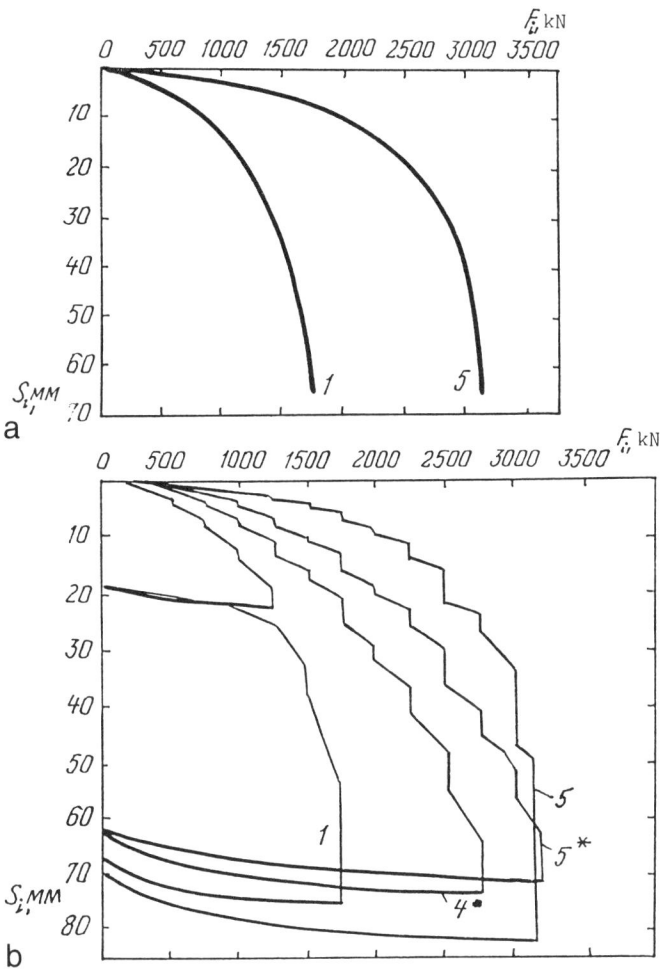

Figure 8.13. Characteristic charts 'load F_i – settlement S_i'; according to the data from pile testing in sand (traditional-technology piles and post-grouted piles): a) typical charts; b) charts borrowed from Stocker's research. 1. without post-grouting; 4*. with an incomplete post-grouting of the foundation; 5*. with the post-grouting of only the skin surface; 5. with the grouting of both the skin surface and the lower end.

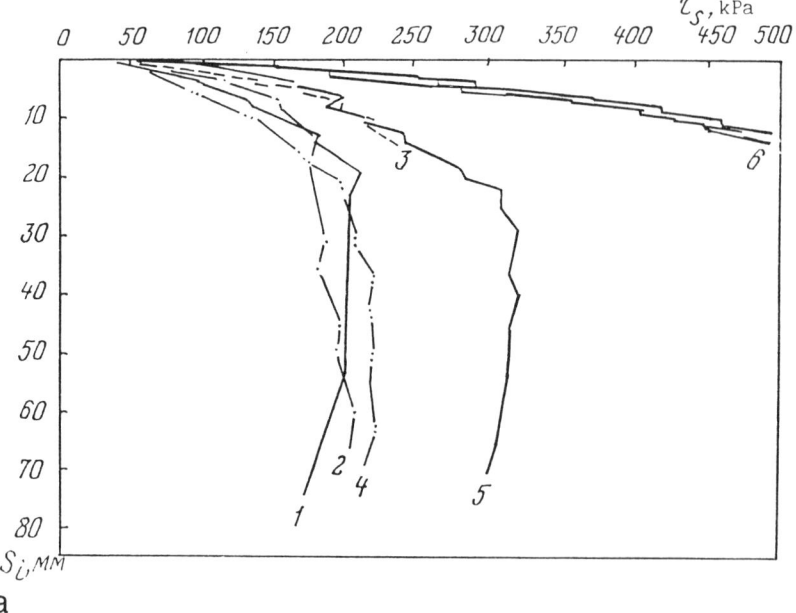

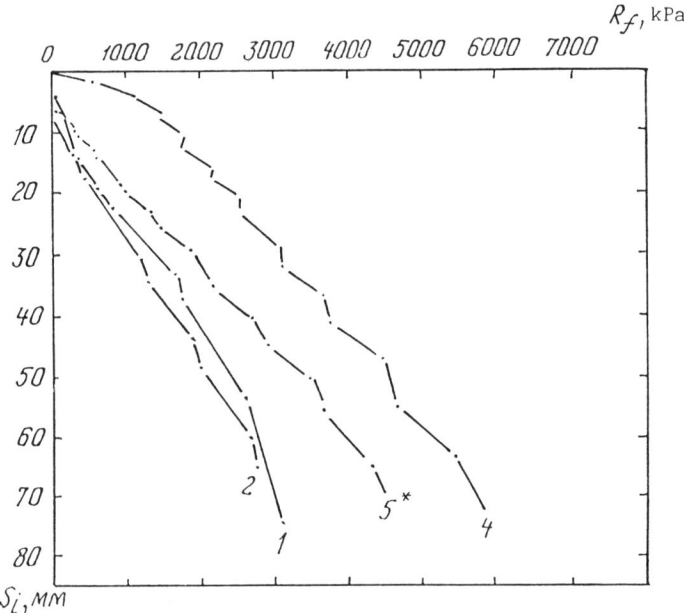

Figure 8.14. Functions: a) 'contact friction τ_s – settlement S_i'; b) 'lower end resistance R_f – settlement S_i' for piles of types 1, 2, 4, 5, 6 (5* – grouting of only the skin surface) according to Stocker.

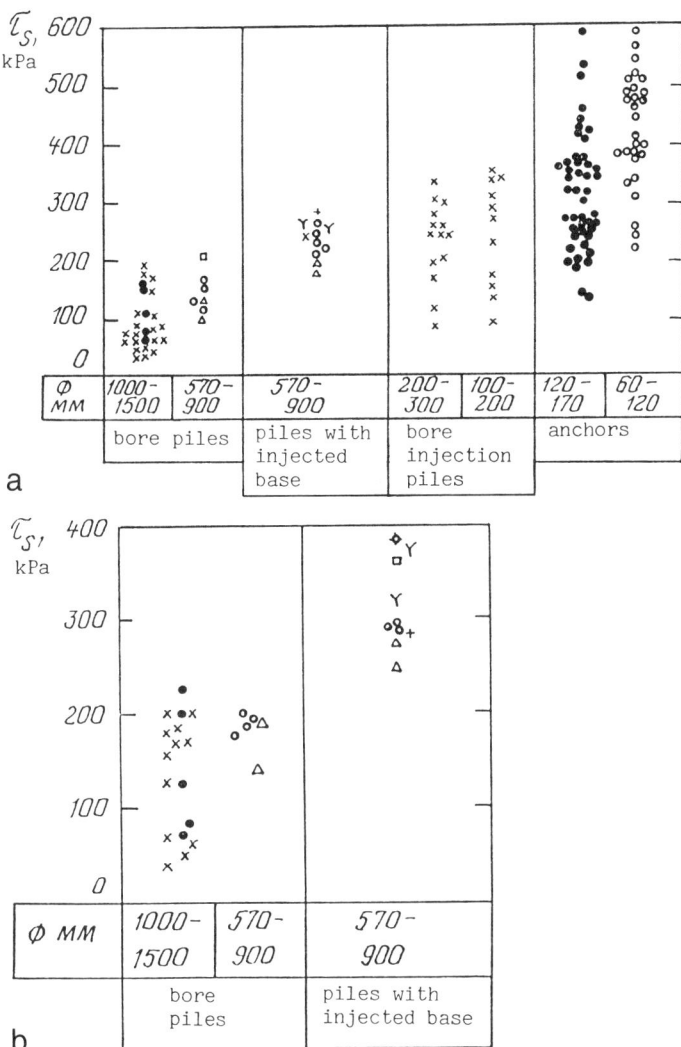

Figure 8.15. Values of contact friction for piles constructed without post-grouting and with injection of the skin surface, and for soil anchors when the settlement equals: a) 10 mm; b) 30 mm.

pile base. Slurry from under the lower end of the pile is completely displaced and the pressure relief at the bottom of the hole is compensated for. Grout forms a coating around the pile which (the coating) is about 50 mm thick, and fills in the possible caverns and empty spaces formed in the process of drilling (Fig. 8.12a, b). As a result of such compacting the surface and the lower end begin to work at commensurable values of settlements (10-20 mm) when the pile is loaded.

Fig. 8.13a, b represent characteristic test charts for piles of Types 4 and 5 in comparison with piles of Type 1. Values of contact friction and resistances under the lower ends of piles of Types 4 and 5 are given in Fig. 8.14a, b.

Fig. 8.15a, b represents the values of τ_s with settlement 10 and 20 mm for large diameter bore piles constructed with and without skin surface injection in comparison with small diameter injection piles and anchors. In plotting the chart we made use of the data provided by Kempfert (1982), Stocker (1980); for injection piles by Stocker & Koreck; for injection anchors by Ostermayer (1974). Elastic deformation of the pile shaft under load was not taken into account.

The charts represent data for small diameter injection piles (diameter 200-300 mm) constructed with excavation performed by rotatory drilling accompanied with water or bentonite suspension flush, and by air flush of the mortar under the pressure of 5 to 6 bar. 100 to 200 mm diameter piles were constructed using a technology analogous to the one used for anchors with grouting through a borehole casing while it is being extracted.

The data from the mentioned research works testify to the increase of τ_s with the decrease of the diameter of the shearing body. This agrees with the theoretical research by Meissner (1982), Wernick (1979). The latter made an attempt to connect this phenomenon with the increasing influence of dilatancy. Stocker (1983) justly adds that the loosening of soil, diminishing the bearing capacity, increases when the hole diameter is increased.

The chart in Fig. 8.16 represent the values of resistances under the lower end of bore piles with the diameter of 570 to 1500 mm R_f (settlement 10, 20 and 30 mm). Its analysis reveals a quasi-linear increase of R_f with the settlement and grain size.

A nice illustration of this is also provided by the charts of functions $\tau_{s_i} = f(S_i)$

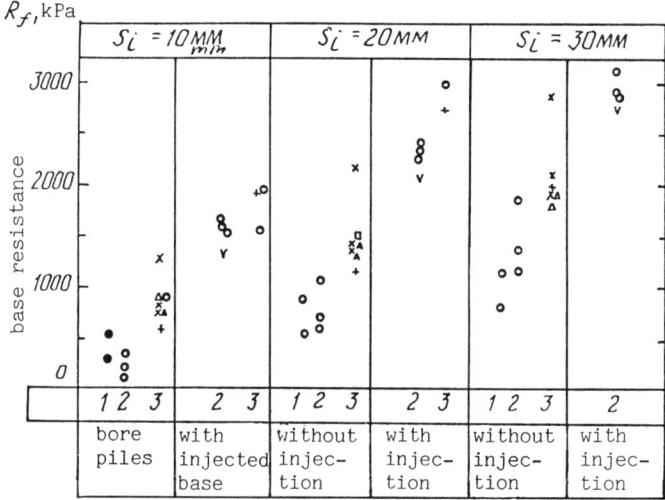

Figure 8.16. Resistances of sandy soil under the lower ends of piles constructed without post-grouting and with post-grouting, mobilised when settlement S_i = 10, 20 and 30 mm. 1. mean density sand; 2. dense sand; 3. gravel from medium to dense (Stocker, 1983).

and $R_{f_i} = f(S_i)$ plotted as a result of a special comparative study of piles of various types where pressures were measured by pressure gauges built into the pile bodies (Fig.8.14).

It follows from the chart in Fig.8.14 that piles of Types 1, 2, 4 with non-compressed skin surface have approximately the same contact friction. In the case of piles with post-grouting there is an increment of by 40-60%. Close values of τ_s were also registered in the case of piles of Type 3 constructed with the use of auger technology, i.e. when concrete is fed under pressure. On the contrary, the friction along the skin surface of piles of Type 1 is considerably lower. Consequently, lateral pressure of concrete which has undergone vibration is not capable alone to compensate for the loosening of soil.

High values of τ_s for injection piles of Type 6 with the diameter of 135 mm are also noteworthy. When the settlement equals 10 mm they are almost 100% higher than those measured for post-grouted piles with the diameter of 570 mm. It is clear that the values of τ_s are influenced not only by contact layer soil properties determined by the technology, but also by the diameter of the shearing body. This factor has not as yet been given a satisfactory explanation in the body of literature on the problem, but it is fairly well understood in the light of the dilatant approach set forth above (see Sections 2.2 and 7.2.2).

Fig. 8.14b gives experimental values of resistances at the lower ends of piles from the quoted research by Stocker. Piles of Types 1, 2, 5 with non-grouted base display the same behavioural pattern. The maximum pressure increases approximately linearly with the settlement. A certain initial settlement (5-8 mm) is needed, though, to mobilise the base resistance. A totally different dependence is observed in the case of piles of Type 4 with additional post-grouting. The resistance rapidly rises to 1,000-1,500 kPa, after which the curve runs parallel to the curves plotted for piles constructed without grouting.

8.6 INJECTION PILES AND ANCHORS

Injection anchors in the modern understanding of the term are known since 1958, when they were used by the German firm 'Bauer Schrobenhausen' for supporting the Munich TV Center excavation pit (Hanna, 1982). Later the injection anchoring technology was employed for constructing small diameter piles which found ample use in reconstruction works.

The sequence for constructing an anchor and piles referred to by us as Type 6 (Table 8.1) is represented in Figs 8.17 and 8.18. The technology is based on injecting cement grout into holes with the diameter of 80 to 150 mm (more rarely up to 200 mm), which allows the achievement of fixing of the tie or of the reinforcing element. The grout pumped into the hole is usually a liquid water-cement suspension with the component ratio from 0.4 to 0.6.

There are two main injection methods. In the first method the grouting is

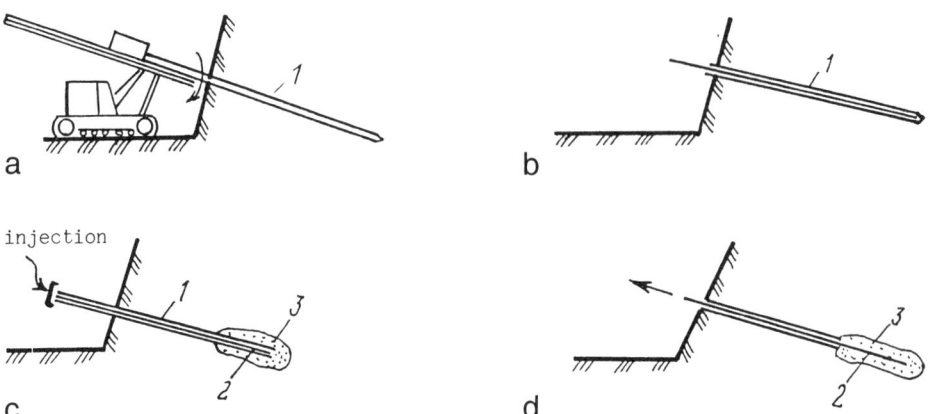

Figure 8.17. Sequence for constructing an injection anchor: a) hole drilling; b) formation of the root by injection of cement grout; c) pre-stressing of the anchor. 1. hole; 2. tie; 3. root.

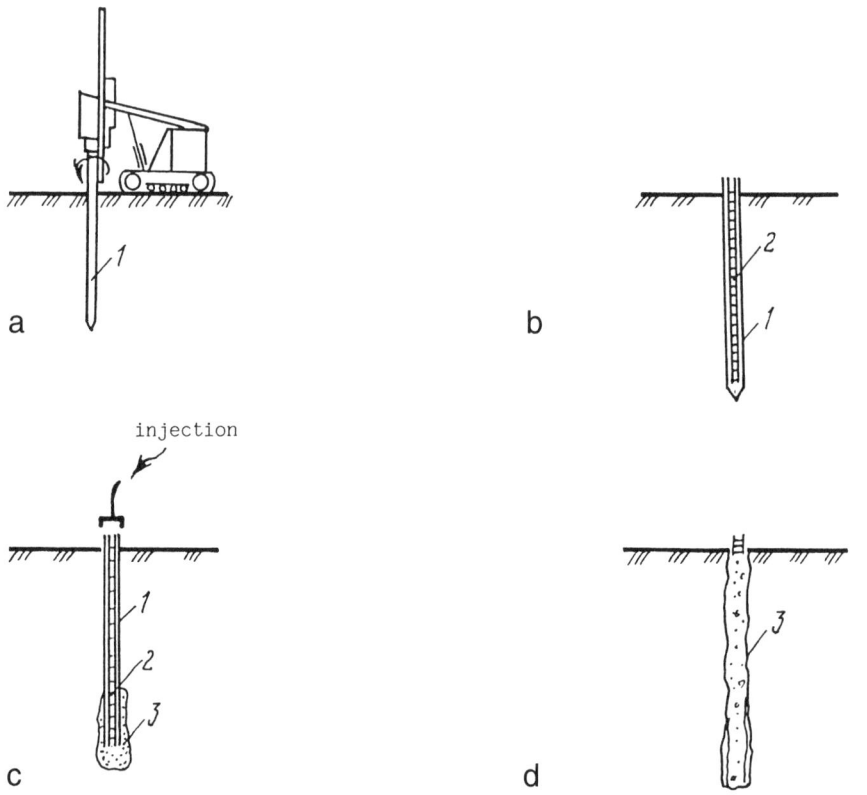

Figure 8.18. Sequence for constructing an injection pile: a) hole drilling; b) placement of a reinforcing bar, cage or profile; c) formation of the pile body by cement grout injection; d) constructed pile. 1. hole; 2. pile reinforcement; 3. injected pile shaft.

Figure 8.19. Cement bodies obtained as a result of injection in sands: a) in homogenous soil; b) at the border of layers with different density and permeability (Sobolevsky, 1985).

effected directly through the borehole casing, and the injection body is formed while the casing is being gradually extracted. According to the second method, the plugging of the hole and filling it in with mortar are effected first, then 24 to 48 hours later injection is effected through a special valved pipe. The technologies of anchoring and their peculiarities were studied by Hanna (1982); Hobst & Zajic (1977, 1983); Schnabel (1982); Sobolevsky (1985).

When the injected grout comes into contact with non-cohesive soil, an intensive 'wringing' takes place, and the original water-cement ratio rapidly drops to 0.25 to 0.27. The 'wringing' almost instantly lowers the mortar consistence and limits its infiltration into the soil pores cementing them in a 20 to 50 mm thick layer. The thrust of the hole by the grout pressure together with the filtrating flux of the displaced water create conditions for a considerable compaction of soil in the area around the pile. Owing to such compaction the initial hole diameter can increase by 2 to 4 times.

Fig. 8.19 represents cement bodies formed after injection into a 114 mm diameter hole in the sand of average density. A detailed study of forming injected bodies was made by Sobolevsky (1985).

The compacting of soil upon contact with the injected body creates ideal conditions for the realisation of the constrained dilatancy factor. The result is very high values of contact resistance to shear.

According to a full-scale research by Ostermayer & Sheele (1977), the ultimate contact friction in the case of anchors reaches the values of 150 to 300 kPa in the sands of originally loose and medium density, and up to 800 kPa in very dense sands. This fully agrees with the data obtained by Koreck (1976) for injection piles, and with the results of our research (Sobolevsky & Nikitenko 1980-1985, Sobolevsky 1984-1988, Sobolevsky & Popov 1987-1988). The values of contact friction for injection anchors and piles based on the data of experiments conducted by various authors are also represented in Fig. 8.15 (Stocker, 1983).

The first attempt to explain such high values of contact friction by the effect of dilatancy was probably made by Wernick (1977). Analyzing the interrelation of contact friction and normal pressure on the surface of anchor root he noted that the values of angles of internal friction do not correspond to those received in the course of tests conducted with the use of tri-axial and shear apparatuses. Wernick attempted to explain this contradiction, frequently encountered in the analysis of the bearing capacity of deep foundations, by the fact that dilatancy causes an increase of angles of internal friction as a result of mutual gear and wedging of grains (see Section 1.7). Wernick also came to the important conclusion that the resistance of soil to shear in the area contacting with the root is not a natural pressure function, but is determined by dilatancy.

8.7 PILES CONSTRUCTED WITH SOIL DISPLACEMENT

These piles referred to as Type 7 in Table 8.1 were the most widely spread in the previously USSR. Accordingly, most of the research conducted in this country was dedicated to them. We shall limit ourselves to a brief summary of those peculiarities which have a bearing on the further analyzed dilatant approach.

The driving of a pile is accompanied by displacement of soil into the surrounding massif. Such displacement is possible due to the compaction in a certain plastic zone around the pile and elastic compression of the massif outside this zone (Fig. 8.20).

According to our notions, the plastic deformation of shear along the pile shaft takes place under the conditions of constrained dilatancy. The contact grain layer is limited with the pile shaft on one side and by the soil massif on the other side. Evidently, the reaction of the massif to thrust on the side of the circular shear zone determines the pressure generated along the pile shaft when it is being loaded. Conditions for the realisation of constrained dilatancy are created due to the plastic deformations of compaction in the process of soil displacement. This can be interpreted as a preliminary spring compression in the model represented in Fig. 2.1.

An attempt to solve this problem of displacement in the soil of a cylindrical hole, and also the problem of the pressure established on the skin surface of a driven pile was made be Lapshin (1987). The problem was solved on the basis of

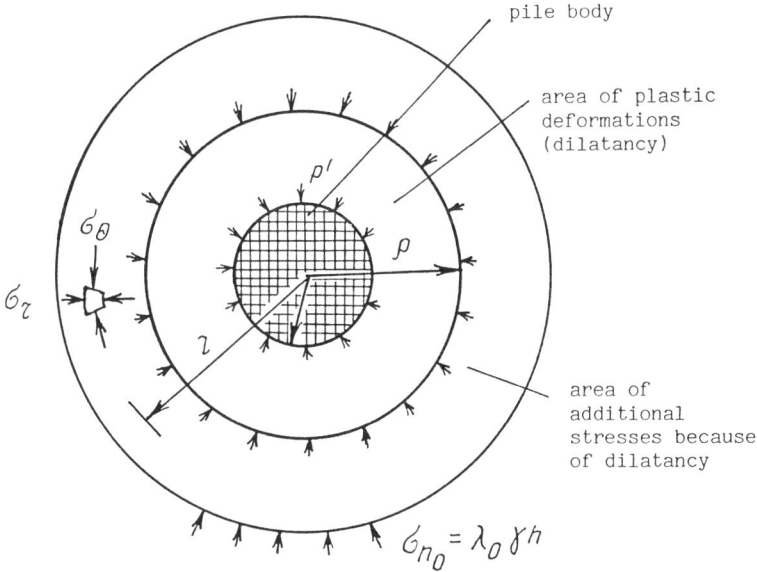

Figure 8.20. Deformation areas and stress condition around the pile constructed with soil displacement (p – thrust pressure of displacement).

an elasto-plastic model, as for widening of the hole in a certain layer from zero to the cross section of the introduced body.

Lapshin jointly considered the conditions of limiting equilibrium of the axis-symmetric plane deformation in the cylindrical system of coordinates, and the Coulomb-Mohr strength condition. Let it be remembered that taking into account the correction for dilatancy introduced here (see Section 7.3) the latter has the following form for non-cohesive soil:

$$\frac{\sigma_r - \sigma_\theta}{\sigma_r + \sigma_\theta + 2\tau_d \operatorname{ctg} \varphi'} = \sin\varphi'. \tag{8.3}$$

Lapshin (1987) deduced formulas for determining the initial pressure under which a zone of limiting equilibrium P_{in} is formed, the radius of the plastic deformation zone ρ, and the residual lateral pressure of the pile shaft P'. This approach is undoubtedly justified, especially for cohesive plastic soils when the pile is introduced without considerable dynamic influences and vibration. We shall attempt to employ it for non-cohesive soil taking into account the strength conditions of constrained dilatancy.

The initial pressure with specific cohesion $C = 0$, corresponding to the formation of a plastic area constitutes

$$P_{in} = \gamma h(1 - \sin\varphi_0) \tag{8.4}$$

For cohesive soil

$$P_{in} = P_0(1 + \sin\varphi) + C\cos\varphi \tag{8.5}$$

where P_0 is natural horizontal pressure.

If we take into account dilatancy, which is analogous to substitution of C for τ_d in the equation deduced by Lapshin for cohesive soil,

$$P_{in} = P_0(1 + \sin\varphi') + \tau_d \cos\varphi'. \tag{8.6}$$

A comparison of the values on P_{in} received from (8.4) and (8.6) is given in Table 8.6. The values of φ_0, φ' and τ_d were taken for medium-density sands of control mixes (Section 4.1) according to Tables 5.1 and A1.8-A1.12.

As follows from Table 8.6 the values of P_{in} allowing for dilatancy turn out much higher than those expected according to (8.4). However, it is necessary to keep in mind that they correspond to the passive thrust of the massif to the static widening of the hole. That is why their use appears to be possible only for 'pressed-in' piles with a reservation that additional corrections will be made by the contact dilatancy along the shaft of the introduced body.

In the case of ramming or vibro-immersion of a pile, a considerable decrease in the values of τ_d and φ'_c takes place, because the mutual re-packing of grains is facilitated due to a decrease in their gear. Internal friction is largely determined by accelerations imparted to the grainy medium. That is why the answer to the problem of initial and residual pressure of soil on the pile shaft should take into

Table 8.6. Values of initial pressures of formation of the plastic deformation zone calculated according to (8.4) and (8.6).

Pile diameter d, m	Natural horizontal pressure P_o, kPa	Values of P_{in} (kPa) in sand					
		Coarse		Medium		Fine	
		acc. to (8.6)	acc. to (8.4)	acc. to (8.6)	acc. to (8.4)	acc. to (8.6)	acc. to (8.4)
0.2	20	142	18	78	22	74	23
	40	172	36	107	45	104	46
	60	202	54	137	67	133	70
0.4	20	116	18	67	22	64	23
	40	147	36	97	45	94	46
	60	178	54	127	67	123	70

account the dynamics of the soil massif.

In the analysis of the stress condition around the pile after its driving, special attention is given to the relaxation of lateral pressure. Allowing for the dilatant approach, the relaxation can also probably be explained as a result of post-packing of grains in the plastic deformation zone.

During the driving of the pile the original package of grains is distorted in a certain area limited by the radius ρ (Fig. 8.20). Part of the displaced grains do not have time to immediately occupy the position of stable mutual balance. Over-stressing leading to crushing in contact points may appear on their edges. A certain time is needed for post-packing and micro-shears; during this time a certain decrease in dilatancy and tangential extension pressures takes place. This is reflected in the corresponding decrease of elastic compression pressures in the surrounding massif, i.e. on relaxation.

The lateral pressure on the pile shaft is important for defining the value of contact resistance to shear τ_s, but, taking into account the complex processes during the pile-driving, its exact calculation is difficult. If the calculation of τ_s is attempted according to the traditional reading of the Coulomb law in the form

$$\tau_s = P' \operatorname{tg} \varphi_0 \tag{8.7}$$

the defined contact friction primarily depends on the settled lateral pressure on the pile shaft P'. The impossibility of the theoretical calculation is often explained by the difficulty of its definition.

But if we introduce corrections for dilatancy and define τ_s according to (7.1) as

$$\tau_s = P' \operatorname{tg} \varphi'_c + \tau_d \tag{8.8}$$

P' becomes a quantity of the second order, because friction to a certain degree depends on the value of the dilatant component τ_d.

138 *Constrained dilatancy as a factor of load-holding capacity*

This is possibly the main reason for the fact that beginning with a certain depth (3-4 m) the values of contact friction down along the pile shaft measured during special tests remain almost constant.

If we proceed from this assumption it appears possible to suggest that, while there are no precise methods, theoretical calculation of τ_s be made proceeding from the initial lateral pressure in the massif in the form

$$\tau_s = \lambda_0 \gamma h \mathrm{tg} \varphi'_c + \tau_d \qquad (8.9)$$

where λ_0 – coefficient of horizontal pressure in the quiescent state. The increment of horizontal pressure due to soil displacement should then be referred to the safety margin.

The driving of a pile depends on the resistance of soil to bulge from under its lower end. Theoretically this problem is solved on the basis of strength conditions with internal bulge in Section 9.2.2. The application of this solution is also admissible only for static immersion, while dynamic and vibratory influences require separate treatment and allowance.

8.8 FACTORS OF STRESS-DEFORMATIVE CONDITION AT THE CONTOUR OF A BORE PILE

Upon construction of a hole without casing, the stability of its walls is ensured by tangential dilatant thrust and capillary forces. The condition of the soil can be represented in the form of two zones; a contour zone of plastic and elasto-plastic deformations around the hole, and a zone of elastic deformations in the massif (Fig. 8.21).

According to Nadai (1969), plastic deformations of the contour zone, proceeding from the Mohr theory, can be interpreted as a family of sliding lines in the form of logarhythmic spirals (Fig. 8.22).

The formation of a hole is accompanied by a local elastic relief of the massif. The elastic displacements occurring in the process of relief are small and do not exceed the values of limiting dilatant strains. Upon unloading the massif radial pressure is established on the outer border of the contour zone. This pressure must probably correspond to the lateral pressure in the minimal limit state, which is expressed in the case of constrained dilatancy according to (7.9) as

$$P_r = \sigma_\theta \, \mathrm{tg}^2(45° - \varphi'/2) - 2\tau_d \, \mathrm{tg}\,(45° - \varphi'/2) \qquad (8.10)$$

where $\sigma_\theta = \lambda_o \gamma h$ is the lateral pressure of the soil in the considered section of the hole-length in the quiescent state.

The stresses of elastic relief and the lateral pressure are transformed in the contour zone into stresses of tangential thrust. These stresses are dilatant by their nature because they are conditioned by the mutual wedging of the grains – the arch effect. It is due to this effect that the contour zone performs the role

Factors of stress-deformative condition at the contour of a bore pile 139

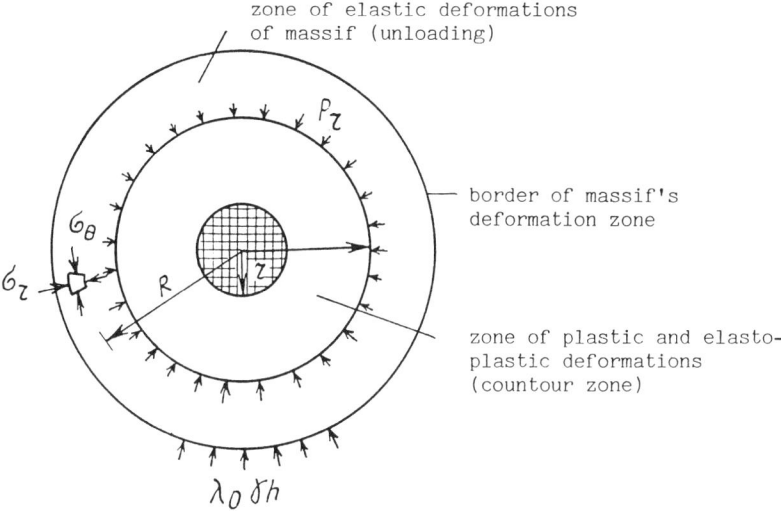

Figure 8.21. Stress-deformative condition of the soil around a cylindrical hole.

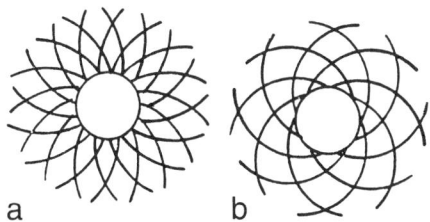

Figure 8.22. Sliding lines around a cylindrical orifice in the free-flowing medium (Nadai, 1969): a) with thrust; b) with unloading.

analogous to that of the casing. Such phenomena are absent in the case of flat walls of a trench in the soil, because there are no conditions for mutual wedging of the grains.

Dilatant stresses for a cylindrical hole in non-cohesive soil serve as a stability reserve analogous to cohesion. They, jointly with capillary forces, prevent development of plastic deformations in the contour zone and are the reason for stability of deep small-diameter holes in dense sands in the absence of any support. The drying up of the soil accompanied by the removal of capillary forces rapidly leads first to local, and then to general failure. Therefore, it is very efficient to periodically pour clayey or cement-clayey mortar into the hole while drilling (Nikitenko & Sobolevsky, 1989). It is rubbed into the hole's walls by the drilling organ and thus creates a thin layer preventing the soil from drying up quickly.

As it is shown above, the values of dilatant stresses can be rather high, especially when the radius of the failure zone is not large (see Tables

A1.7-A1.12). This accounts for the high stability of holes with the diameter of 200-400 mm in dilatant sands. In the case of large-diameter holes the width of the zone where the natural soil structure is violated increases, and as a result of its decompaction in the process of drilling the possibility for the realisation of a dilatant thrust decreases. The role played by the dilatant stresses can be regarded as analogous to that of cohesion: $\sigma_d = \tau_d \operatorname{ctg} \varphi'$. The vertical stresses in this case can also be substituted by the weight of equivalent soil layer with the height

$$h_{eq} = \frac{\tau_d \operatorname{ctg} \varphi'}{\gamma} \tag{8.11}$$

where γ is the unit weight of soil.

When borehole casings are employed, the stresses influencing their walls can change within a wide range depending on the drilling method and the degree of deformation of contact zone. In the case of ram-cable drilling these stresses can be virtually absent in the places of cavern formation, whereas when the borehole casing is pushed in or immersed into the soil massif by vibration it can approach the lateral pressure in the soil or even exceed it due to the displacement of soil and contact dilatancy. If the drilling method does not entail a forced soil displacement, we can observe the above-mentioned unloading of the massif with formation of a loosened contour zone accompanied by an indefinite stress-deformative condition. In the case of drilling under the protection of clayey suspension the stability is ensured by thrust pressure of liquid column h_i

$$P_r = \gamma_s h_i$$

where γ_s is the unit weight of the suspension.

The filtration of the suspension through the hole's walls is accompanied by their quick filling with clay particles (see Section 8.3). This rapidly decreases the consumption of filtrating water and ensures the conditions for the appearance of hydrodynamic forces which are an important factor of stability. Let us clarify this with an example.

Suppose the thickness of the clayey layer is $\delta_c = 2$ mm, the head of the liquid in the hole $h_1 = 6$ m and on the side of ground waters level $h_2 = 4$ m. The hydraulic gradient

$$i = \frac{h_1 - h_2}{\delta_c} = \frac{6 - 4}{2 \times 10^{-3}} = 1000 \tag{8.13}$$

The corresponding volume filtration force with the unit weight of water $\gamma_w = 10$ kN/m^3

$$f = \gamma_w i = 10^4 \text{ kN/m}^3 \tag{8.14}$$

The drilling under the protection of the clayey suspension ensures a better preservation of the natural soil structure around the hole in comparison with the other technologies. The action of hydrodynamic forces and hydrostatic pressure

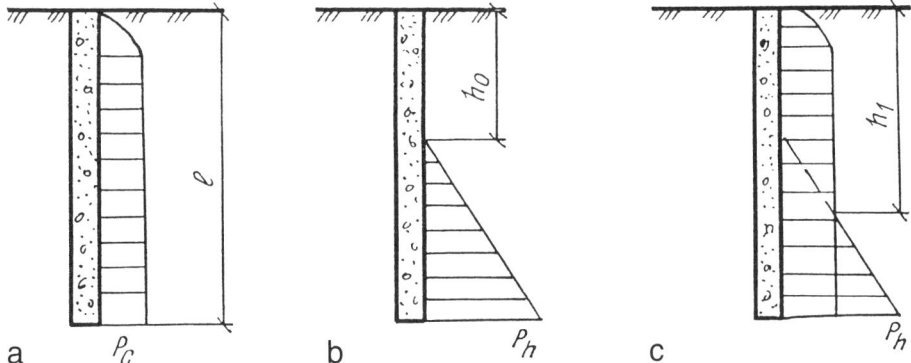

Figure 8.23. Orthographic epure projections of radial pressures along the shaft of a bore pile (Lapshin, 1987): a) pressure of concrete mix on the walls; b) pressure of soil along the contour of the conrete-filled hole; c) pressure of soil on the pile shaft (resulting).

inhibits the elastic relief of the massif and the dilatancy in the contour zone. The border of the latter approaches the hole's walls.

The stress-deformative condition of soil around a bore pile is finally determined after its concreting. The pressure of the fresh concrete on the hole's walls depends mainly on the method of its placement, the rigidity of the mixture, grain size of the aggregate and the shrinkage. In the case of compaction of the concrete by vibration its lateral pressure can be assumed to correspond to hydrostatic pressure:

$$P_c = \gamma_c h_c \tag{8.15}$$

But the height of the concrete column h_c here corresponds only to the height of the vibrated section, because above it the concrete is capable of partially hanging on to the walls of the hole and transmitting its weight lower down not in full measure.

As Lapshin (1987) pointed out, it can be assumed that when $h_c \geq 2d$ the pressure on the concrete mix remains constant all along the hole length and with the unit weight of concrete $\gamma_c = 23$ kN/m³ (Fig. 8.23) constitutes

$$P_c \approx 8.211d \tag{8.16}$$

Horizontally the pressure P on the skin surface of a bore pile is determined either by the pressure P_c generated during the filling in of the hole with concrete allowing for its shrinkage upon setting, or by the horizontal pressure P_h at the considered depth (Lapshin, 1987), i.e.

$$P = P_c \text{ when } P_c \geq P_h \tag{8.17a}$$

$$P = P_h \text{ when } P_h > P_c \tag{8.17b}$$

142 *Constrained dilatancy as a factor of load-holding capacity*

The latter condition ensures joint work of the pile and the massif, i.e. the condition of constrained dilatancy under shear.

The condition (8.17a) presupposes either localisation of shear deformations in the loosened contour zone, or partial reaction of the massif to dilatancy of the contact layer after a certain initial settling.

One should not confuse the pressure on the wall of the hole with the pressure at the outer circle of the contour dilatant zone. The former can change from zero to P_c with the filling in of the hole with concrete, while the latter determines the degree of inclusion of the massif into work when the pile is loaded.

For analysis we shall use an approach analogous to that suggested by Lapshin (1987). Radial pressures around the hole in dilatant soil at the depth h when the hole is filled in with clayey suspension (unit weight γ_s) taking into account (8.12) will constitute (Spivak & Popov, 1985)

$$\sigma_r = \lambda_o \gamma \left(h + h_{eg} \right) \left(\frac{r^2}{R^2} - 1 \right) - \left(\gamma_s h + \tau_d \operatorname{ctg} \varphi' \right) \frac{r^2}{R^2} \tag{8.18}$$

where r is the radius of the hole;

R is the distance from the axis of the hole to the point in which the pressure is being determined.

At the contour of the hole (with $r = R$)

$$\sigma_r = -\gamma_s h - \tau_d \operatorname{ctg} \varphi' \tag{8.19}$$

For a hole not filled in with liquid

$$\sigma_2 = -\tau_d \operatorname{ctg} \varphi' = -\sigma_d \tag{8.20}$$

The pile will constitute a unified massif with the surrounding soil if the pressure of concrete or of its post-grouting equals or exceeds the horizontal pressure at the outer border of the contour zone. Returning to Fig. 8.22 it can be represented as a reverse turn of the grains along spirals of sliding lines with resumption of their

Table 8.7. Hole depths down to which the pressure of concrete does not compensate the contour dilatancy according to the calculation for control-mix sands.

Hole radius r, m	Values of h_o, m, in sands					
	Coarse		Medium		Fine	
	Dense ($I_D = 1.0$)	Medium density ($I_D = 0.6$)	Dense ($I_D = 1.0$)	Medium density ($I_D = 0.6$)	Dense ($I_D = 1.0$)	Medium density ($I_D = 0.6$)
0.2	52*	26	29	11	23	10
0.4	41	20	23	9	18	8
1.0	30	15	17	6	14	6
2.0	24	12	13	5	11	5
3.0	21	10	12	4	9	4

* Only factor of dilatancy is evaluated theoretically.

initial position, or as dilatancy in the reverse direction with the appearance of an external thrust to the side of the massif.

Assuming (8.20), the horizontal pressure at the contour of a concrete-filled hole is

$$P_h = \lambda_o \gamma h - \tau_d \operatorname{ctg} \varphi'. \tag{8.21}$$

If we assume that transversal widening in the massif is absent, then

$$\lambda_o = \frac{1-\nu}{\nu}. \tag{8.22}$$

Then, proceeding from (8.21) and taking (8.22), the pressure of the soil at the contour of a concrete-filled hole is absent until the depth (Fig. 8.23b)

$$h_o = \frac{\tau_d \operatorname{ctg} \varphi'(1-\nu)}{\nu \gamma}. \tag{8.23}$$

Let us note that if, on the average, $\varphi' = 33°$, $\nu = 0.3$ and $\gamma = 18$ kN/m³, then

$$h_o \approx 0.2\,\tau_d. \tag{8.24}$$

It is of interest to assess the order of the values obtained according to (8.23). Let us draw upon the data from Tables A1.7-A1.12 bearing on the values of τ_d with $E = 356$ MPa. Table 8.7 gives the hole depths corresponding to them, which turn out to be unexpectedly large. Taking them into account, it is possible to conclude that in the majority of cases the thrust of the concrete mixture is insufficient for the compensation of elastic relief of the massif and of the accompanying dilatancy of the contour zone.

Analysis reveals why contour injection with shaft base compression is so efficient, particularly for piles originally constructed using technology of Type 1. With post-grouting under pressure equalling 3-5 MPa, which exceeds the lateral pressure of soil by a factor of 10^2 and the value of dilatant stresses by an order of magnitude, the degree of soil looseness in the contour zone and the value of tangential dilatant thrust becomes irrelevant. The grouting compresses and compacts the soil, compensates the elastic relief of the massif and creates optimum conditions for the realisation of the maximum dilatancy and resistance to shear.

Down to a certain depth h_1 the calculated horizontal pressure on the pile shaft is determined by the condition (8.17a). Lapshin (1987) deduced an expression for h_1 for cohesive soil (Fig. 8.23c). Substituting the stress τ_d (acquired 'dilatant cohesion') for the specific cohesion C, we receive

$$h_1 = \frac{(8.211d + \tau_d \operatorname{ctg} \varphi')(1-\nu)}{\nu \gamma}. \tag{8.25}$$

This, probably, is not strictly correct if the stability prior to the placement of concrete is ensured by drilling mortar or by soil on the auger's blades preventing the development of contour dilatancy around the hole. In this case

$$P_h \approx \lambda_o \gamma h \tag{8.26}$$

the condition (8.17) is met along the major part of the pile length, and the value of h_1 can be calculated if we equate (8.26) and (8.16):

$$h_1 = \frac{8.211 d}{\lambda_o \gamma} \tag{8.27}$$

If we assume that on the average $\lambda_o = 0.38$ and $\gamma = 18$ kN/m^3, then

$$h_1 \approx 1.2 d \tag{8.28}$$

When concrete based on ordinary portland-cements hardens in the conditions where humidity of the environment is less than 90% the shrinkage phenomenon is clearly manifested. With the drop in humidity from 90% to 25% the shrinkage increases by approximately 6-7 times. The value of average annual shrinkage of heavy concretes $\Delta\varepsilon_s = 0.0002\text{-}0.0004$. The shrinkage increases with the rise in the specific content of cement and with the increase of the water-cement ratio of the concrete mix.

The phenomenon of shrinkage of concrete in the hole was studied by Lapshin (1987). His research revealed that shrinkage is manifested when there is a possibility of unrestricted outflow of water from concrete into the soil. The influence of pressure of concrete mix on those processes has not been discovered.

In low-humid sandy soils there are open conditions for the outflow of free water from the concrete mix. They are absent only when the shaft is concreted below the level of ground water and are restricted when the walls of the hole are clay-filled.

The radial deformations of the concrete shaft can be defined by the formula

$$\Delta r = \Delta\varepsilon_s \frac{d}{2} \tag{8.29}$$

If we assume, according to Lapshin (1987), that $\Delta\varepsilon_s = 0.0003$, then with the diameter of the pile equalling, for instance, $d = 0.4$ m the shrinkage will constitute 0.06 mm, and with $d = 1.0$ m it will be 0.15 mm.

The shrinkage is negligibly small in respect to the pile diameter, but if we compare it with the ultimate dilatant strains δ_d (see Table A1.3) it will transpire that these quantities are either of the same order or coincide. And if this is so, then in the light of the dilatant approach shrinkage can cause considerable influence on the stress condition of the contour zone.

Studying the shrinkage process in application to soil anchors we found out that with sands, practically immediately upon filling the hole in with cement mortar without injection, an outflow of water into the soil takes place (mortar with the 0.4-0.6 water-cement ratio was used). At the same time the cement body showed a visible shrinkage of up to 1-2 mm, and the grain layer adjacent to it 'separated' itself from the walls of the hole forming a visible 'clearance' with the surrounding soil. As a result the mobilised values of contact friction decreased and the shear settlement increased (Nikitenko & Sobolevsky, 1986; Sobolevsky, 1985).

The deformations of shrinkage are partly compensated for by the transversal widening of the concrete shaft when the pile is loaded. The radial elastic deformations under the influence of the force F can be defined as

$$u_r = \frac{2F}{\pi d E_c} v_c \qquad (8.30)$$

If the modulus of elasticity of concrete $E_c = 3.0 \times 10^3$ MPa, the Poisson's coefficient $v_c = 0.18$ and $F = 2000$ kN, the u_r with the pile diameter $d = 0.4$m will constitute 0.019mm, while with $d = 1.0$m it will equal 0.0076 mm, i.e. 32% and 20% of the shrinkage Δr calculated above, respectively. Such partial compensation of the shrinkage, though, does not appear sufficiently definitive.

8.9 CONCLUSION: REASONS FOR THE FAILURES OF THEORETICAL CALCULATION METHODS

The low values of contact resistance to shear and high values of settlement for certain types of bore piles are accounted for by the following reasons:
– Loosening of the soil along the walls of the hole and at its bottom during the process of drilling;
– Elastic relief of the massif with the formation of a contour zone of plactic deformations and a rise of the hole's bottom;
– Shrinkage of the concrete and formation of a 'clearance' along the pile shaft.

These reasons introduce vagueness into the assessment of the stress condition of the soil around the pile both prior to and in the process of its loading. The loosening of the soil, and even more so the formation of a 'clearance' along the pile shaft do not ensure the conditions of constrained dilatancy. At the same time the lateral pressure of the soil is transmitted to the pile incompletely, and the contact shear is developed in the weakened soil with diminished values of strength parameters.

In these conditions the soil along the contour of the pile can sustain contraction, i.e. the 'clearance' can still grow further in the presence of shear settlement. This accounts for such low values of calculated resistances along the skin surface of the pile and of work condition coefficients in Tables 2 and 5 of SNIP 2.02.03-85.

When the shear settlement is depleted the forces of resistance to shear along the shaft reach their maximum and further increase of load is possible only at the expense of resistance of foundation under the lower end. Here, however, there is an accumulation of loosened soil, and to compress it, a settlement considerably increasing the shear settlement (30-40 mm and more) may be needed. A still larger settlement is necessary for a complete involvement into work of the soil under the lower end, i.e. for the formation of plastic deformation zones and of an area of internal bulge. As a result, upon reaching by the pile of maximum bearing capacity the summarised settlements may achieve 50, 80 mm and more, that is become unacceptable for the erected structure.

The vagueness of the stress condition of the soil, of its density and strength parameters render the theoretical calculation of a bore pile of Type 1 on the basis of the limit equilibrium theory difficult or impossible. Such vagueness also makes it difficult to introduce corrections for the influence of dilatancy into the calculation.

Technologies of Types 2 and 3, connected with drilling under the protection of clayey suspension or ensuring temporary support owing to the hanging up of the soil column on the auger's blades, have a considerable advantage. Lesser massif relief and loosening of the contact layer give a possibility to carefully employ conditions of strength of the equations (7.1) and (7.7) in the calculations, i.e. to take into account dilatancy when the original density of the surrounding soil exceeds the critical value. These technologies, however, do not completely exclude the decompaction of the soil and the formation of a weakened zone under the lower end of the pile.

Bore piles of Types 5 and 6 display the best results (Table 8.1). During their construction, thrust stresses generated in the process of injection by far exceed the initial pressures of plastic deformations and the stresses of elastic relief. The stresses relieved during the drilling of the hole are returned to the massif. The pile then appears to be jammed in the soil while the natural lateral pressure on its shaft is resumed or even exceeded. This ensures the conditions necessary for the complete realisation of constrained dilatancy.

If the injection is effected with high quality it always results in a considerable compaction of the contact layer of the soil. Research (Sobolevsky, 1985) has revealed that density of sands along the contour of the hole, after grouting with a mortar-cement ratio of about 0.5, becomes close to maximum possible, i.e. tends to the values of I_D close to 1.0. Therefore it appears justified to recommend the definition of strength parameter of the contact layer using the modelling of the injection effect. This possibility is partly ensured by a vent-shear apparatus described by Nikitenko & Sobolevsky (1986).

A preliminary post-compaction of the sample can be effected during testing in DACS-type apparatus (see Section 3.2). The optimum method of such post-compaction is imparting it with vibratory movements accompanied by constant normal pressure. The vibration ensures a compaction effect similar to the effect obtained by the influence on the soil of hydrodynamic forces during injection (Sobolevsky, 1985).

Injection of mortar under the lower end of the pile also ensures displacement and compaction of slurry and loosened soil. At the same time the weakened zone, causing large settlements which precede the involvement into work of the base

Conclusion: Reasons for the failures of theoretical methods 147

Table 8.8. Principal negative factors of technologies for construction of piles, footings and anchors.

No.	Factor	Piles 1	2	3	4	5	6	Footings 1	2	3	Injection anchors	Character of the factor's influence
1	Loosening of soil along the walls of the hole (cutting)	+	+	+	+	−	−	+	+	−	−	Indefenite decrease of the values of strength parameters, diminishing or exclusion of conditions of constrained dilatancy
2	Accumulation of loosened soil or slurry at the bottom of the hole (cutting)	+	+	−	−	−	−	+	−	−	−	Increase of settlement required for the mobilisation of the passive resistance of soil with internal bulge
3	Elastic relief of the mass along the contour of the hole (cutting)	+	−	−	+	−	−	−	−	−	−	Diminishing or exclusion of conditions of constrained dilatancy, increase of shear settlement
4	Elastic relief of the massif at the bottom of the hole (cutting)	+	−	+	−	−	−	−	−	−	−	Increase of the settlement needed for the mobilisation of the passive resistance of soil
5	Formation of a clayey contact layer	−	+	−	−	−	−	+	+	−	−	Vagueness of the values of strength parameters
6	Shrinkage of concrete	+	+	−	+	−	−	+	+	−	−	Decrease of conditions of constrained dilatancy

+ The factor is present or possible.
− The factor is absent or compensated by a special technology.

resistance, is eliminated. Original conditions for the beginning of internal bulge area formation directly after the load application are provided.

As a result of injection the mobilisation of the soil resistance during the loading takes place along the shaft and under the lower end of the pile with commensurable settlements, i.e. the skin surface and the lower end of the pile are involved into work simultaneously. The definiteness in the soil condition after the post-grouting compaction opens up the possibility to use a definite calculation scheme and create a calculation method based on the conditions of strength of dilating soil.

An efficient calculation method must be founded on a calculation scheme which is close to reality and includes reliable soil and construction parameters. Vagueness in the mentioned elements creates indefiniteness in the logical design of the whole calculation method.

Certain peculiarities of some technologies of deep foundation engineering were noted above. The factors exerting negative influence upon the bearing capacity and creating vagueness during calculation are summarised in Table 8.8.

The conclusion inevitably comes to mind that, when we deal with non-cohesive soils, the technologies connected with the forced post-compaction of the soil are actually directed, though not quite consciously, at bringing the factor of constrained dilatancy to work. However, the latter has still not been given the deserved consideration in special literature, even though the measured values of contact friction for piles of Type 5 and 6, and anchors equalling 200, 300 kPa and more directly contradict the Coulomb-Mohr theory.

At the same time the application of the Coulomb law of the form (7.6) in the traditional reading to the bore piles of Types 1 and 2 often yields values of τ_s close to the actual values. This is connected with the absence of diminishing of constrained dilatancy due to the localisation of the shear only in the loosened contact layer. On the basis of the above-mentioned facts it is possible to formulate a number of questions which have not been supplied with adequate answers in the light of modern soil mechanics approaches:

– Due to what are values of contact friction by an order of magnitude greater than those expected according to Coulomb achieved?

– What is the reason for the increase of values of contact friction with the decrease of the diameter of the shearing body?

– How can the values of resistance under the lower ends of piles and footings be theoretically substantiated?

An attempt to answer these questions is contained in Part 1 of this work. The purpose of this chapter was to employ the concept of constrained dilatancy set forth in Part 1 for the analysis of the technological influence on the properties of soil.

In the light of the dilatant approach the technology for deep foundation engineering constitutes a factor determining the conditions for mobilisation of the soil strength which can occur according to the schemes of both constrained and free dilatancy. Traditional technologies of constructing bore piles allow considerable changing of soil which causes the ambiguity of the conditions for the mobilisation of contact friction and resistance to internal bulge. The above-mentioned ambiguity, together with the lack of understanding of peculiarities of free and constrained dilatancy, seriously hampers the creation of a theoretical calculation method or makes it impossible.

The injection technologies ensure the compaction of soil accompanied by compensation for the loosening sustained in the process of excavation, and the mobilisation of resistances to the contact shear and to internal bulge in the conditions of constrained dilatancy with minimum settlements. These technologies open possibilities for direct influence on the soil with an aim to achieve optimum conditions for its work.

Pile construction technologies with the excavation of soil by the hollow-stem auger method, under the protection of bentonite suspension and by diaphragm wall method allow to exclude considerable changes of soil and relief of stresses acting therein. With strict adherence to the requirements of these technologies there is a possibility to ensure conditions of constrained dilatancy, i.e. the optimum work of soil when the foundation is loaded. Thus the factor of constrained dilatancy is the principal reserve for enhancing the bearing capacity and increasing the efficiency of deep foundations. The realisation of this fact opens the way to improving the existing and working out new technologies ensuring active and purposeful influence on the soil.

CHAPTER 9

Load-holding capacity of a single pile

9.1 CALCULATION SCHEME

The load-holding capacity of a pile is made up of the forces of resistance to contact shear along the skin surface F_s and to pressing under the lower end F_f, i.e.

$$F_p = F_s + F_f \tag{9.1}$$

Fig. 9.1a represents the design scheme for a single pile, Fig. 9.1b – the scheme for a bore pile with injected (grouted) base and skin surface. The following principal dimensions can be distinguished: length of the pile L; length of the non-injected section L_1; length of the grouted section L_2; pile's diameter d; diameter of the injected section d_2; and (for piles with square section) side length b.

An assumption has been accepted regarding an even distribution of mean contact resistance to shear $\bar{\tau}_{si}$ within each homogenous soil layer. While establishing the reaction of massif to dilatancy, modulus of elasticity and Poisson coefficient are assumed to be mean for the intersected soil layers with the thickness l_i

$$\bar{E} = \frac{\Sigma E_i l_i}{L}, \quad \bar{v} = \frac{\Sigma v_i l_i}{L} \tag{9.2}$$

and coefficient of elastic resistance is assumed to correspond to the whole area of contact between the pile surface and soil, i.e.

$$K = \frac{2\bar{E}}{(1 + \bar{v})d} \tag{9.3}$$

9.2 PROPOSITIONS REGARDING CALCULATION OF BORE PILES

9.2.1 *Load-holding capacity along the skin surface*

The load-holding capacity of a pile along its skin surface can be expressed as

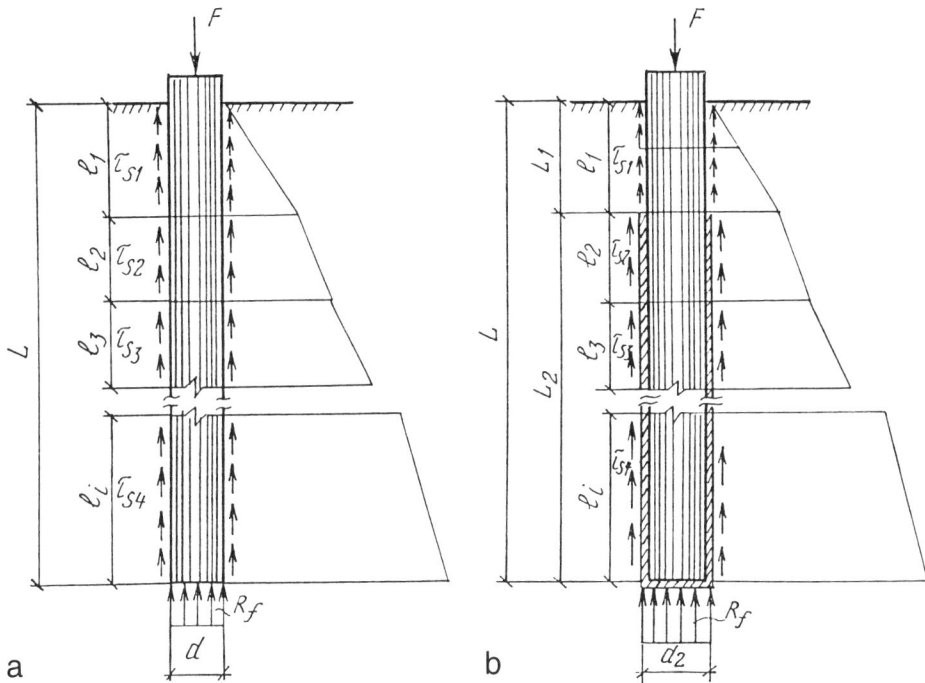

Figure 9.1. Pile design scheme: a) bore pile or pile manufactured with soil displacement; b) bore pile with injected lower end and (partially) shaft.

$$F_s = \pi d \sum \bar{\tau}_{si} l_i \qquad (9.4)$$

Fig. 9.1 represents stresses active along the skin surface of a pile. Prior to loading it is the lateral pressure of soil taken as corresponding to the quiescent state

$$\sigma_{s_o} = \lambda_o \sum \gamma_i l_i \qquad (9.5)$$

where λ_o is the coefficient of lateral pressure of the i-th soil layer in quiescent state; γ_i is the unit weight of soil of the i-th soil layer taken to be equal to the natural (initial) weight.

For piles manufactured in holes under the temporary protection of bentonite suspension (Type 2, see Table 8.1) or by feeding mortar through a hollow-stem auger (Type 3), the value of $\bar{\tau}_{si}$ is calculated taking into account the conditions of constrained dilatancy according to (7.1). Then

$$F_s = \pi d \sum \bar{\tau}_{si} l_i = \pi d \sum \left(\lambda_{oi} \gamma_i l_i \, \text{tg} \, \varphi'_{ci} + \tau_{di} \right) l_i \qquad (9.6)$$

where the pile diameter is assumed to be equal to the hole diameter. Strength parameters τ_{di} and φ'_{ci} are calculated respective to initial density of soil.

In a way of example, Table A2.1 represents calculated values of mean contact resistances to shear $\bar{\tau}_s$ for homogenous control-mix sands (Section 4.1a), while in

Table 9.1. Values of contact resistance to shear for bore piles of types 2, 3.

Calculated values according to (7.2)		According to Stocker (1983)	
Pile diameter d, m	$\bar{\tau}_s$, kPa	Pile dimater d, m	$\bar{\tau}_s$, kPa
0.50...1.00	82...204	0.57...0.90	90...210
1.00...1.50	78...197	1.00...1.50	35...190

Table 9.1 they are compared with experimental data brought together by Kempfert (1982) and Stocker (1983) (Fig. 8.15). The calculated values of $\bar{\tau}_s$ were taken for the pile length 10 m.

In the diagram in Fig. 8.15, one can observe an increase of contact resistance to shear with the decrease of pile diameter. Stocker (1983) points out that it can be partially accounted for by the increasing effect of dilatancy with lessening of the diameter. He refers to the corroborating research by Meissner (1982) and Wernick (1979), though he mentions that the noted influence of the pile diameter can also be explained by the differences in the pile manufacturing method.

9.2.2 Load-holding capacity at the lower end

Soil failure at the lower end of the pile takes place in the conditions of internal bulge. The formation of the bulge area is preceded by compaction deformation of soil and the possible remains of the slurry. These deformations can be quite considerable and exceed the acceptable limit for the given erecting structure.

Let's consider the conditions of soil work under the lower end of the pile after the compaction deformations have been exhausted. As Lapshin points out (1987), development of deformation zones under the lower ends of the piles upon their loading occurs, generally, sideways.

Figs 9.2 and 9.3 represent the patterns of soil-grain movements under the lower end of a model pile and the deformations accompanying them (according to Lapshin). At first a wedge shaped compacted nucleus is formed (Fig. 9.2a); upon its submergence into the massif it pushes it aside, 'cleaves' it (Figs 9.2b and 9.3) and forms an oval area of plastic deformations.

An isochrome pattern obtained by Lapshin (1987) in his experiments by the photoelasticity method is represented in Fig. 9.4. It is interesting in that it characterized the development of the maximal tangent stresses under the heel of the pile and gives a qualitative picture of possible deformative development. A major conclusion arrived at by Lapshin is the fact that the form of the pile end does not practically influes on load-holding capacity and that soil plastic deformations spread down under the heel of the pile to the depth in the order of two pile diameters.

Baholdin (1986) in his research on driven piles shows that:

a) Pile settlement from static load takes place mainly sideways at the expense

Propositions regarding calculation of bore piles 153

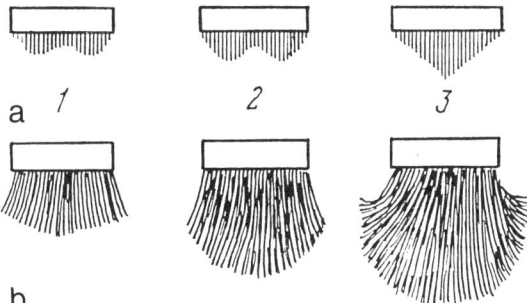

Figure 9.2. Formation of a bulge area under the heel of the pile (model experiments of Lapshin, 1987): a) snapshots made by a moving camera; b) snapshots made by a stationary camera. Settlements: 1. 20 mm; 2. 40 mm; 3. 60 mm.

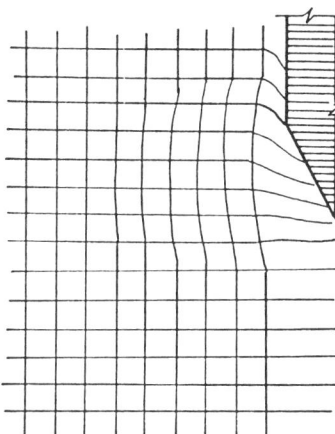

Figure 9.3. Deformation sheme for vertical and horizontal bands during pile submerging (Lapshin's experiments, 1987).

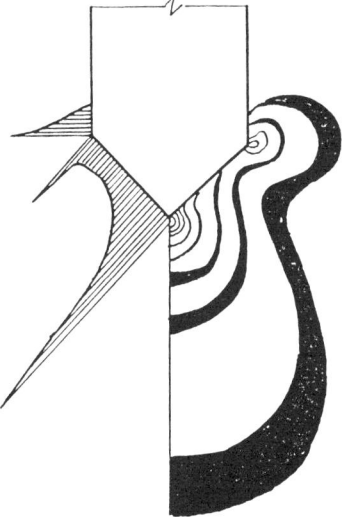

Figure 9.4. Epure projection of maximum tangent stresses (left) and normal stresses (right) for a model pile (photoelasticity method of Lapshin, 1987).

154 Load-holding capacity of a single pile

of soil compaction in the process of an almost horizontal soil bulge from under the heel of the pile;

b) Horizontal internal bulge of soil from static load is localized in a small layer of soil adjacent to the heel of the pile; the thickness of this layer equals about two sides of the pile's cross section;

c) Stresses appearing around piles during soil bulge from static load are several times higher than stresses acting in other directions.

Fig. 9.5 features conditions of limiting equilibrium of soil in lengthwise sections. The failure area is represented as the top wedge AOA and two side quasi-elastic wedges AOB and A′OB′ which thrust the massif aside horizontally.

Vertical stresses acting on the pile-heel level are represented by the pressure of the soil's own weight $\Sigma \gamma_i l_i$ and by additional pressure due to the friction along the skin surface of the pile σ_{ad}. The values of these stresses developing when a pile is loaded were established by Dalmatov & Lapshin (1975), Lapshin (1987) on the basis of the Patie solution (1963) as

$$\sigma_{ad} = \bar{\tau}_s \beta \qquad (9.7)$$

where β is the coefficient taking into account the ratio of the pile-heel submerging to its radius $L/r(2L/d)$ and the distance from its axis to the point of the massif considered in the given case.

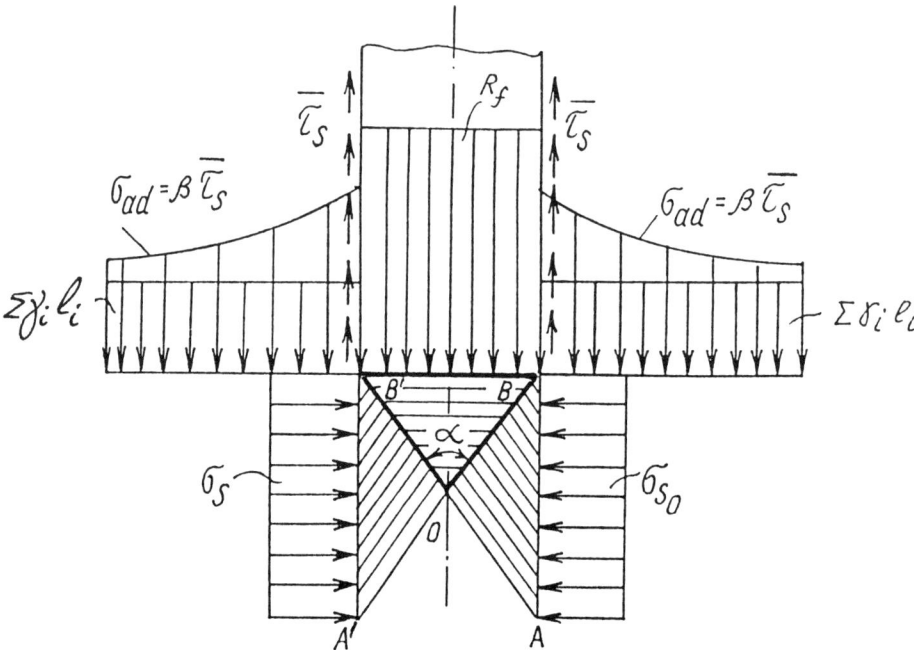

Figure 9.5. Conditions of limiting equilibrium of soil under the lower end of pile (calculation scheme).

Table 9.2. Values of coefficients β (Lapshin, 1987).

L/r	Ratio R/r				
	1	3	5	7	10
5	1.2416	0.2136	0.0799	0.0319	0.0088
10	1.2568	0.2550	0.1349	0.0807	0.0393
20	1.2601	0.2640	0.1520	0.1043	0.0669
30	1.2608	0.2654	0.1544	0.1084	0.0735

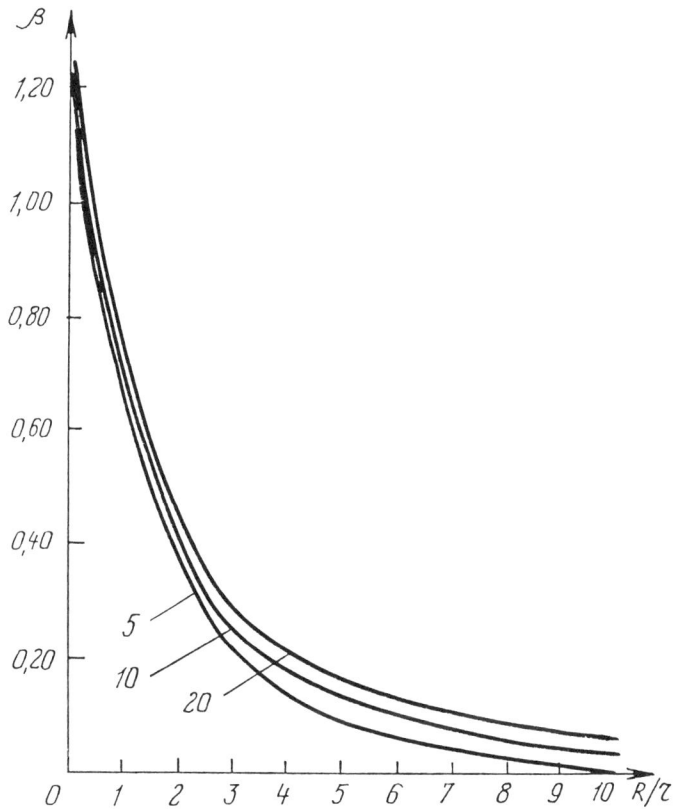

Figure 9.6. Values of coefficient β depending on the relative distance to the pile shaft with $L/r = 5$, 10 and 20 (Lapshin, 1987).

The values of coefficient β, with the value of coefficient of cross expansion of soil $\nu = 0.3$, are given in Table 9.2.

It is seen from Table 9.2 and Fig. 9.6 that the largest additional stresses develop along the pile contour, where they constitute about 1.24-1.26 of the mean shearing stress. They rapidly decrease when the distance from it increases, and with $R/r = 3$ they do not already exceed 0.21-0.26 $\bar{\tau}_s$.

156 *Load-holding capacity of a single pile*

The question regarding the necessity of taking these stresses into account remains disputable, but they are undoubtedly capable of exerting a certain influence on the process of internal bulge due to an increase of resistance to the grain movement sideways and from under the pile heel. That is why we will allow for these stresses. In calculation examples in Table A2.1, the value of thea $\bar{\sigma}_{ad}$ the pile contour is assumed to be equal to $1.26\bar{\tau}_s$.

The equilibrium of the ring wedge-form area in Fig. 9.5 is ensured when there is an equality of the forces of pressing

$$F_f = R_f A_f \tag{9.8}$$

where R_f is resistance at the lower end of the pile; A_f is the area of pile section; and forces of lateral thrust

$$F_s = \sigma_s A_s \tag{9.9}$$

where A_s is the area of the lateral surface of a cylinder with the height AB = kd, then

$$A_s = k\pi d^2 \tag{9.10}$$

From the condition of equilibrium

$$R_f A_f = \sigma_s A_s \tag{9.11}$$

or, taking into account (9.8) and (9.10),

$$F_f = k\pi d^2 \sigma_s \tag{9.12}$$

It can be concluded that the solution of the sum about ultimate soil resistance can be reduced to establishing the maximus resistance of the massif to the expanding of a cylindrical hole with diameter d.

Then,

$$\sigma_s = \lambda_{oi}\left(\Sigma\gamma_i l_i + \beta\bar{\tau}_s\right)\lambda_{pi}^2 + 2\tau_d\lambda_{pi}, \tag{9.13}$$

where

$$\lambda_{pi} = \text{tg}\left(45° + \frac{\varphi'_i}{2}\right)$$

φ'_i is the angle of internal friction of soil.

The ultimate load-holding capacity of a pile at its lower end can be expressed by the equation

$$F_f = k\pi d^2\left[\lambda_{oi}\lambda_{pi}^2\left(\Sigma\gamma_i l_i + \beta\bar{\tau}_s\right) + 2\lambda_{pi}\tau_{di}\right] \tag{9.14}$$

If we assume, allowing for the data obtained by Baholdin (1986) and others, that the internal bulge area spreads under the pile heel for approximately two diameters, the $k = 2$ and, consequently,

$$R_f = \sigma_s \frac{A_s}{A_f} \approx 8\sigma_s \qquad (9.15)$$

Calculated resistances of soil to internal bulge for control-mix dense ($I_D = 1.0$) and medium-dense ($I_D = 0.6$) sands are represented in Table A2.2. It can be inferred from these data that density and grain size of the soil have a major influence on the values of R_f, whereas the pile diameter can be classified as a minor factor. The decrease of coefficient of elastic resistance of massif K with the increase of the pile diameter is compensated for by a large support area. As a result, the largest deviations of mean calculated values of R_f for piles with the diameter of 0.5-1.5 m do not exceed 5%.

Mean calculated values of R_f are also represented in Table A2.2. It is curious to compare them with the values fixed in SNIP 2.02.03-85 'Pile foundations' based on a limit state design. Such comparison is given in Table 9.3. Calculated values of R_f are given for sands from medium-dense to dense, while SNIP represents the values recommended for medium-dense sands.

As it turns out, calculated values of R_f for medium-dense sands differ from those given by SNIP maximally by 22% for coarse sand and 20% for medium sand, and for certain pile lengths the two sets of values almost completely coincide. A greater divergence of values can be observed for fine sand. In this connection we must remind the reader that researched sands formed from only one natural.

Mobilization of ultimate resistance at the lower end of a bore pile occurs with a certain settlement s_i. This settlement can considerably exceed the acceptable s_{max}, at which only part of the potential load-holding capacity is mobilized. This can be allowed for by introducing into (9.14) a coefficient of acceptable settlement n, i.e.

$$F_f = nk\pi d^2 \left[\lambda_{oi} \lambda_{pi}^2 \left(\Sigma \gamma_i l_i + \beta \bar{\tau}_s \right) + 2\lambda_{pi} \tau_{di} \right] \qquad (9.16)$$

where n is the coefficient of settlement acceptable for the given erecting structure.

Table 9.3. Resistances under the lower ends of piles in sandy soils of medium density according to theoretical calculations and SNIP 2.02.03-85.

Soil	Values of R_f, kPa, with pile length											
	6		8		10		12		14		16	
	Theor.	SNIP	Theor.	SNIP	Theor.	SNIP	Theor.	SNIP	Theor.	SNIP	Theor.	SNIP
Coarse sand	5550-9225	7150	6340-10020	7435	7030-10790	7700	7820-11610	7900	8720-12425	8100	9495-13280	8260
Medium sand	2815-5110	3550	3410-5750	3800	3920-6730	4000	4570-7020	4160	5140-7650	4320	5600-8255	4480
Fine sand	2700-4120	2300	3270-4690	2470	3830-5263	2600	4385-3830	2720	4950-6400	2840	5505-6995	2960

158 *Load-holding capacity of a single pile*

If in the first approximation we assume a linear functional relationship of F_f to settlement, coefficient n can be expressed as

$$n = \frac{S_i}{S_{max}} \tag{9.17}$$

Such treatment agrees with proposals of the European Pile Committee (ISSMFE TC3) according to which one should differentiate between true and conditional ultimate load-holding capacity of a pile. Baholdin (1986) treated the true holding capacity as the ultimate load under which, as a result of complete internal bulge of soil, the pile settlement continuously increases, while conditional holding capacity is understood as the load corresponding to the value of stabilized settlement.

The values of strength parameters for calculation according to (9.16) should be established as corresponding to the natural density of soil composition (see Section 8.9).

9.2.3 *Expression for calculating the total load-holding capacity*

Taking into account (9.6) and (9.16), we propose the following expression for the load-holding capacity of a single bore pile of Type 2 or 3:

$$F = \gamma_c \{ \gamma_{cs} \pi d \Sigma (\lambda_o \gamma_i l_i \, \text{tg} \, \varphi'_{ci} + \tau_{di}) l_i$$
$$+ \gamma_{CR} n k \pi d^2 [\lambda_{oi} \lambda_{pi}^2 (\Sigma \gamma_i l_i + \beta \bar{\tau}_s) + 2\lambda_{pi} \tau_{di}] \} \tag{9.18}$$

where γ_c is coefficient of work conditions of pile in the soil; γ_{cs}, γ_{CR} are, respectively, coefficients of work conditions of soil along the skin surface and at the lower end of the pile.

9.3 PROPOSITIONS FOR CALCULATING BORE PILES WITH INJECTED BASE AND SHAFT

9.3.1 *Load-holding capacity along skin surface*

The bearing capacity of piles with injected shaft (referred to Type 5 in our classification, see Table 8.1) can be expressed as

$$F_s = F_{s1} + F_{s2} = \pi (d l_1 \bar{\tau}_{s1} + d_2 l_2 \bar{\tau}_{s2}) l_i \tag{9.19}$$

where F_{s1}, F_{s2} are, respectively, load-holding capacities of non-injected and injected pile sections; $\bar{\tau}_{s1}$, $\bar{\tau}_{s2}$ are respectively mean contact resistances to shear along the non-injected and grouted pile sections.

The conditions of shear of soil at the upper non-injected section of pile shaft depend on the hole-drilling and concreting technologies. For technologies of Type 2 and 3, which ensure the conditions of constrained dilatancy, the holding capacity

F_{s1} should be calculated according to expression (9.6). As for the manufacture of the upper section of the pile with the use of technology of Type 1 the contact resistance to shear is proposed to be calculated as for the conditions of free dilatancy (7.6). Then

$$F_{s1} = \pi d \Sigma (\lambda_{oi} \gamma_i l_i \, \text{tg} \, \varphi_{oi}) l_i \qquad (9.20)$$

where d is the diameter of the non-injected section of the pile shaft, assumed to be equal to the hole diameter; φ_{oi} is the angle of contact friction under condition of free dilatancy.

The value of φ_{oi}, strictly speaking, should be assumed as corresponding to the soil density in the contact layer. This latter quality is hard to establish, though, which is why it seems optimal to introduce an adjustment for the decrease of density directly into the value of φ_{oi}, decreasing it by 1 to 3°. Such approach is well interpreted by the Hansen-Landbourne formula (1.5).

Contact shear along the lower injected section of the pile shaft develops in a layer, compacted by pumping cement grout. This fact ensures conditions of constrained dilatancy and high values of resistance to shear (see 7 Section 8.8). The load-holding capacity can be calculated as

$$F_{s2} = \pi d_2 \Sigma (\lambda_{oi} \gamma_i l_i \, \text{tg} \, \varphi'_{ci} + \tau_{di}) l_i \qquad (9.21)$$

where φ'_{ci} is the angle of contact friction of the soil layer compacted by injection.

The mean diameter of the lower section of the pile shaft can be approximately calculated from the volume of the pumped mortar, taking into account its water-cement ratio, from the equation suggested by Mishakov (1980), as

$$d_2 = \sqrt{d^2 + \frac{1.27V(1 + 0.172m)}{(1 + nm)l_2}} \qquad (9.22)$$

where V is the volume of the pumped mortar; m is the ratio of unit weight of cement and water; n is the water-cement ratio of the injection mortar.

The factors determining the process and results of grouting are treated in detail in Cambefort (1970), Sobolevsky (1985).

Theoretically calculated values of mean resistance to contact shear for dense sands are given in Table A2.3. The density index of sands I_D was assumed to be equal to 1.0 and 0.8. The length of the upper non-injected section is taken to be equal to 4.0 m. Let us mention here that the listed calculations are given by way of example and correspond exactly to control-mix sands.

A comparison of the data of Table A2.3 with experimental values of $\bar{\tau}_s$ obtained during in-situ-tests and generalized by Stocker (1983) is given in Table 9.4. The calculated values of lateral friction are assumed to be corresponding to piles with the length of 10 m; the least of them correspond to fine medium-dense sand and piles with diameter of 1.0 m, while the largest correspond to coarse sand and piles with diameter of 0.5 m. Contact resistance to shear for piles with an injected shaft in comparison with traditional bore piles are also given in Fig. 8.15.

160 *Load-holding capacity of a single pile*

Table 9.4. Calculated values of contact resistance to shear for bore piles with injected shaft.

Calculated values according to (7.2)		According to Stocker (1983)	
Pile diameter d, m	$\bar{\tau}_s$, kPa	Pile diameter d, m	$\bar{\tau}_s$, kPa
0.50-1.00	142-283	0.57-0.90	175-285

9.3.2 *Load-holding capacity at the lower end*

For calculating the resistance to internal bulge for a pile with soil compaction by grouting the same calculation scheme as that for usual bore piles can be used (Fig. 9.5). The values of additional stresses on the level of the lower end of the pile are established according to (9.7) with the following difference: the value of mean tangent stresses is taken only for the length of the injected section and is designated as $\bar{\tau}_{s2}$.

Compaction of soil under the lower end of the pile takes the settlement which is necessary to finally compact the loosened layer and the possible remains of the slurry before the resistance of the base to internal bulge is mobilized (see Section 8.9). We therefore suggest, that calculation should be made by expression (9.14), but without allowing for the coefficient of accepted settlement. The values of resistance of soil to internal bulge, for pile diameters equalling 0.5, 1.0 and 1.5 m in control-mix sands, calculated according to (9.14) and (7.10), are given in Table A2.4. Let it be remembered that an injection, when performed correctly, is capable of ensuring soil compaction close to maximally dense composition ($I_D \approx 1.0$). An exclusion can be constituted by water-saturated soils and soils with high-filtrating layers and bulky inclusions (Cambefort, 1970; Sobolevsky, 1985).

9.3.3 *Expression for calculating the total load-holding capacity*

The bearing capacity of piles with an injected heel and shaft (Type 5) when holes are manufactured by technologies of Types 2 and 3 (see Table 8.1) can be expressed proceeding equations (9.12) and (9.14) as

$$F = \gamma_c \{\gamma_{cs1} \pi d \Sigma (\lambda_{oi} \gamma_i l_i \, \text{tg} \, \varphi'_{ci} + \tau_{di}) l_i + \gamma_{cs2} \pi d_2 \Sigma (\lambda_{oi} \gamma_i l_i \varphi'_{ci} + \tau_{di}) l_i$$

$$+ \gamma_{CR} k \pi d_2^2 [\lambda_{oi} \lambda_{pi}^2 (\Sigma \gamma_i l_i + \beta \bar{\tau}_{s2}) + 2\lambda_{pi} \tau_{di}] \quad (9.23)$$

When the upper (non-injected) section of the shaft is manufactured with the use of technology of Type 1, the bearing capacity of the pile, taking into account (9.20) will be

$$F = \gamma_c \{\gamma_{cs1} \pi d \Sigma (\gamma_i l_i \, \text{tg} \, \varphi_o) l_i + \gamma_{cs2} \pi d_2 \Sigma (\lambda_{oi} \gamma_i l_i \, \text{tg} \, \varphi'_{ci} + \tau_{di}) l_i$$

$$+ \gamma_{CR} k \pi d_2^2 [\lambda_{oi} \lambda_{pi}^2 (\Sigma \gamma_i l_i + \beta \bar{\tau}_{s2}) + 2\lambda_p \tau_{di}]\} \quad (9.24)$$

where γ_c is the coefficient of work conditions of pile in the soil; γ_{cs1}, γ_{cs2} are, respectively, coefficients of work conditions of soil along the skin surface of the non-injected and injected sections; γ_{CR} is the coefficient of work conditions of soil under the lower end.

For piles of Type 4, i.e. those manufactured without injection of skin surface, the bearing capacity is expressed according to (9.18), but with the value of coefficient of accepted settlement $n = 1$.

9.4 PROPOSITIONS FOR CALCULATING INJECTION PILES

The bearing capacity of an injection pile (Type 6, see Table 8.1) is primarily determined by the value of resistance to contact shear along the skin surface. The resistance under the lower end of the pile is a minor factor, which is accounted for by the small (relative to the skin surface) area of pile support. Thus, for a pile with diameter 150 mm and length 6 m, the section area constitutes only 0.6% of the skin surface. It is recommended to take the resistance under the lower end of the pile into consideration only when the pile is supported by very dense and rocky soils, i.e. when failure due to the loss of stability is possible (Koreck, 1976).

The following expression is suggested for the load-holding capacity of a bore-injection pile:

$$F = \gamma_c \gamma_{cs} \left[\pi d_2 \Sigma \left(\lambda_{oi} \gamma_i l_i \, \text{tg} \, \varphi'_c \right) l_i + \tau_{di} \right] \qquad (9.25)$$

where φ'_{ci}, τ_{di} are strength parameters of the compacted contact layer; d_2 is the mean pile diameter widened by injection (Fig. 9.1b).

We recommend calculating the value of d_2 by employing formula (9.22).

Examples of mean resistances to contact shear for bore-injection piles in dilating non-cohesive soils are given in Table A2.5. They are calculated according to (7.2) for control-mix sands with $I_D = 1.0$ and 0.8. If we compare Tables A2.3 and A2.5, it turns out that the values of $\bar{\tau}_s$ are considerably higher in the case of bore-injected piles. One should also bear in mind that while manufacturing injection piles of small diameter it is much easier to ensure an effective compaction of soil in the area surrounding the pile. A comparison with the data provided

Table 9.5. Calculated values of contact resistance to shear of injection piles for control-mix sands.

Values calculated according to (7.2)		According to Stocker (1983)	
Pile diameter d, m	$\bar{\tau}_s$, kPa	Pile diameter d, m	$\bar{\tau}_s$, kPa
0.15	113...306	0.10...0.20	100...360
0.20	105...232	0.20...	
0.25	88...262	...0.30	85...340

162 *Load-holding capacity of a single pile*

by Stocker (1983) shows that the calculated values of $\bar{\tau}_s$ obtained by us are quite close to those measured in natural conditions (Table 9.5).

9.5 PROPOSITIONS FOR CALCULATING PILES MANUFACTURED WITH DISPLACEMENT OF SOIL

9.5.1 *Load-holding capacity along the skin surface*

The load-holding capacity of a pile is determined by the value of contact resistance to shear taking place in the conditions of constrained dilatancy. According to the calculation scheme in Fig. 9.1a,

$$F_s = u\Sigma(\lambda_{oi}\gamma_i l_i \text{ tg } \varphi'_{ci} + \tau_{di})l_i \qquad (9.26)$$

where u is the perimeter of the pile skin surface; γ_i is the unit weight of soil of natural composition; φ'_{ci}, τ_{di} are, respectively, the angle of contact friction and the dilatant component of strength of the soil layer of the contact zone.

For round-section piles

$$u = \pi d \qquad (9.27)$$

for square-section piles

$$u = 4b \qquad (9.28)$$

The stress condition of soil around a square-section pile corresponds mostly to the axis-symmetrical sum, and it is for such sums that we recommend the establishment of the parameters φ'_{ci} and τ_{di}. When the coefficient of elastic resistance is established for laboratory soil tests or for calculations, it is recommended to employ the given pile diameter

$$d'_u = 1.274b \qquad (9.29)$$

Correspondingly, the given value of coefficient of elastic resistance of the massif for the skin surface is

$$K'_u = \frac{\bar{E}}{(1 + \bar{v})0.637b} \qquad (9.30)$$

Taking into account (9.28) we write down:

$$F_s = \pi d'_u \Sigma(\lambda_{oi}\gamma_i l_i \text{ tg } \varphi'_{ci} + \bar{\tau}_{di})l_i \qquad (9.31)$$

where for round-section piles $d'_u = d$.

While establishing resistance to shear for the soil surrounding a driven pile, it is extremely important to allow for the influence exerted by the phenomenon of relaxation. This phenomenon can be interpreted on the basis of dilatant notions as a result of final grain compaction, affecting those grains which in the process of pile driving occupied uncertain mutual positions.

Relaxation of contact presupposes a certain additional deformation of compression around the piles. Such deformation can be expressed in a grain repacking in the dilating contact layer with a certain 'unloading' of the dilatant thrust.

Consequent to the general fading away of shear movements, the grains which found themselves in a state of unstable mutual equilibrium tend to acquire a more stable, 'energetically profitable' position. Another possibility is a rupture of those grains on whose facets there develop stresses reaching the values of the mineral's strength (see Section 4.8). The deformations connected with this, according to the model adopted in Section 2.2, mean a decrease of the dilatant component of normal pressure in expression (4.3).

It follows from Section 4.7 that absolute values of dilatant movements δ_d are not great and usually constitute from 1/10 to 1/100 mm. That is why additional deformation from the final grainpacking can turn out to be quite substantial. Let us clarify this proposition with an example.

For instance, as a consequence of relaxation, the movement $\Delta\delta_d$ during shear in medium-sized medium-dense sand, which originally constituted 0.04 mm (with $K = 2845$ MN/m³), decreased by 35% (see Table A1.3). According to (2.1), dilatant stress

$$\Delta\sigma_d = K\Delta\delta_d = 2845 \times 0.04 \times 10^{-3} = 0.114 \text{ MPa} = 114 \text{ kPa}$$

corresponds to the value $\Delta\delta_d = 0.04$ mm. With initial normal pressure on the pile shaft $\sigma_{n_o} = 100$ kPa the corresponding ultimate resistance to shear is

$$\tau_u = (100 + 114) \text{ tg } 30° = 124 \text{ kPa}$$

After relaxation with $\Delta\delta_d$ decreased by 25% $\Delta\sigma_d = 85$ kPa, and

$$\tau_u = (100 + 85) \text{ tg } 30° = 105 \text{ kPa}$$

i.e. less by 15%

9.5.2 *Load-holding capacity at the lower end*

The load-holding capacity of a driven pile at the lower end was established from the conditions of internal bulge according the calculation scheme in Fig. 9.5 (Lapshin, 1987). These conditions are similar for both round and square piles and correspond to an axis-symmetrical sum. The values of R_f given below are established as corresponding to section areas: for a round pile

$$A = \frac{\pi d^2}{4} \tag{9.32}$$

for a square pile

164 *Load-holding capacity of a single pile*

$$A = b^2 \tag{9.33}$$

From the condition of equality of (9.32) and (9.33) the given diameter for the lower end of a square pile is

$$d'_A = 1,128b. \tag{9.34}$$

Correspondingly, the given value of the coefficient of elastic resistance for the lower end of the pile is

$$K'_A = \frac{\overline{E}}{(1-\overline{v})0,564b}$$

Taking into account (9.14),

$$F_f = k\pi d'^2_A \left[\lambda_{oi}\lambda_p^2 \left(\Sigma\gamma_i l_i + \beta\overline{\tau}_s\right) + 2\lambda_{pi}\tau_{di}\right] \tag{9.36}$$

Where for round section $d'_A = d$. We suggest that additional stress on the level of the lower end $\sigma_{ad} = \beta\overline{\tau}_s$ could be taken from the mean shearing stress along the length of the pile.

Table A2.6 gives, as an example, resistances under the lower ends of driven piles in control-mix sands, calculated theoretically according to expressions (9.14) and (7.10) with $k = 2$; while in Table 9.6 we give their comparisons with the values taken from Table 1 of SNIP 2.02.03-85 'Pile foundations'. Table A2.7 represents calculated values of load-holding capacity at the lower ends of driven piles.

9.5.3 *Expression for calculating the total load-holding capacity*

Taking into account (9.29) and (9.34), the bearing capacity of a driven pile is expressed as

$$F = \gamma_c \left\{ \gamma_{cs} \pi d'_{u} \Sigma \left(\lambda_{oi}\gamma_i l_i \text{ tg } \varphi'_{ci} + \tau_{di}\right) l_i \right.$$
$$\left. + \gamma_{CR} k\pi d'^2_A \left[\Sigma\lambda_{oi}\lambda_{pi}^2 \left(\gamma_i l_i + \beta\overline{\tau}_s\right) + 2\lambda_{pi}\tau_{di}\right]\right\} \tag{9.37}$$

Table 9.6. Calculated resistances under the lower ends of driven piles in sandy soils of medium density according to theoretical calculations and SNIP 2.02.03-85.

Soil	Values of R_f, kPa, with pile length											
	6		8		10		12		14		16	
	Theor.	SNIP	Theor.	SNIP	Theor.	SNIP	Theor.	SNIP	Theor.	SNIP	Theor.	SNIP
Coarse sand	5730-7675	7150	6215-8150	7435	6695-8600	7700	7180-9060	7900	7665-9540	8100	8145-10010	8260
Medium sand	3010-3863	3550	3435-4275	3800	3845-4680	4000	4280-5120	4160	4700-5485	4320	5125-5905	4480
Fine sand	2725-3775	2300	3115-4175	2470	3504-4565	2600	3903-4900	2720	4295-5360	2840	4670-5760	2960

9.6 CONCLUSION

Drawing the conclusions from Chapter 9, it is worthwhile reminding the readers once more, that in Part 2 of this work the author did not try to work out refined calculation methods, but only wished to single out and give foundations to possible directions and approaches to the use of refined conditions of strength by Coulomb-Mohr for establishing the load-holding capacity of certain constructions in soil.

The use of refined conditions of strength makes convergence possible of the theoretically calculated and experimental values of load-holding capacity for piles manufactured with the use of different technologies. Allowing for dilatancy makes it possible to provide a theoretical foundation for the dependence of contact resistance to shear along the pile shaft on its diameter.

The revealed functional relationship between the strength of non-cohesive soil and its deformativity opens up possibilities for the optimization of dimensions and construction technologies for piles. At the same time the solving of the problem concerning internal bulge allows the suggestion of a theoretical method for calculating resistance at the lower end of the pile.

Technological advantages condition the best qualities for piles manufactured with injection compaction of the base and for bore-injection piles. Injection compaction of soil allows us to ensure conditions of constrained dilatancy during shear along the pile shaft and to increase contact friction due to this for up to 5 or more times. Elimination by means of injection compaction of the layer under the lower end of the pile, which is loosened in the process of boring, ensures this layer's involvement into work with settlements commensurable with the shearing settlement.

A major advantage of piles manufactured with minimal soil violation or with injection compensation of these violations is the fact that their work is subordinated to a rather precise calculation scheme.

A number of possible directions for further research can be singled out.

1. Refinement of deformative-stress condition of soil around the pile with an account for the technology of its manufacture under the viewpoint of dilatancy.

2. Detailed study of the process of internal bulge in soils of various density. Clarifying the problem of taking into account additional normal stresses from friction along the pile shaft.

3. Refinement of work conditions for piles in a cluster on the basis of taking into account the influence of the summary field of dilatant stresses on the massif. Comparing the bearing capacity of single piles and that of piles in a cluster provides a possibility for theoretically justifying the optimum distance between piles.

4. Studying of relaxation processes whilst allowing for the role of dilatancy, especially for piles manufactured with soil displacement.

5. Generalization of the experimental data on different types of piles and their

analysis with a view to establishing justified values of coefficients of work conditions, acceptable settlement and reliability.

6. Studying the role of dilatancy in resistance to contact shear of piles working for pulling-out, and its (dilatancy's) influence on the volume of the body of soil adhering to the pile.

CHAPTER 10

Load-holding capacity of a deep footing

10.1 CALCULATING SCHEME

Design scheme of a footing (trench foundation) is represented in Fig. 10.1. Its load-holding capacity on soil F is made up of the forces of resistance to contact shear along the skin surface F_s and pressing in under the lower end F_f, i.e.

$$F = F_s + F_f \tag{10.1}$$

The following principal geometrical dimensions of a trench foundation can be discerned: total height H, length L, breadth b. The area of skin surface is designated A, of the end surface – A'. During injection of parts of skin and end surfaces the non-injected (H_1) and injected (H_2) heights are also discerned. Correspondingly, the area of skin surface falls into A_1 and A_2, and the area of end surface is divided into A'_1 and A'_2. The increase of breadth due to injection is not taken into account, because its influence on the total area is very insignificant. The area of the foundation's sole is designated A_f. We have also assumed that mean contact resistance to shear τ_s is evenly distributed within one soil layer along skin and end surfaces. The height of each homogenous soil layer is designated h_i.

While establishing the reaction of the massif to dilatancy, modulus of elasticity and Poisson coefficient are assumed to be mean for the crossed soil layers:

$$\bar{E} = \frac{\Sigma E_i h_i}{H}, \quad \bar{v} = \frac{\Sigma v_i h_i}{H} \tag{10.2}$$

while the coefficient of elastic resistance is assumed to be corresponding to the whole contact area, i.e.

$$\bar{K} = \frac{\bar{E}}{(1 - \bar{v}^2)\omega L}, \quad \bar{K}' = \frac{\bar{E}}{(1 - \bar{v}^2)\omega b} \tag{10.3}$$

10.2 LOAD-HOLDING CAPACITY ALONG THE SKIN SURFACE

The load-holding capacity of a footing along the skin surface is determined by the

168 Load-holding capacity of a deep footing

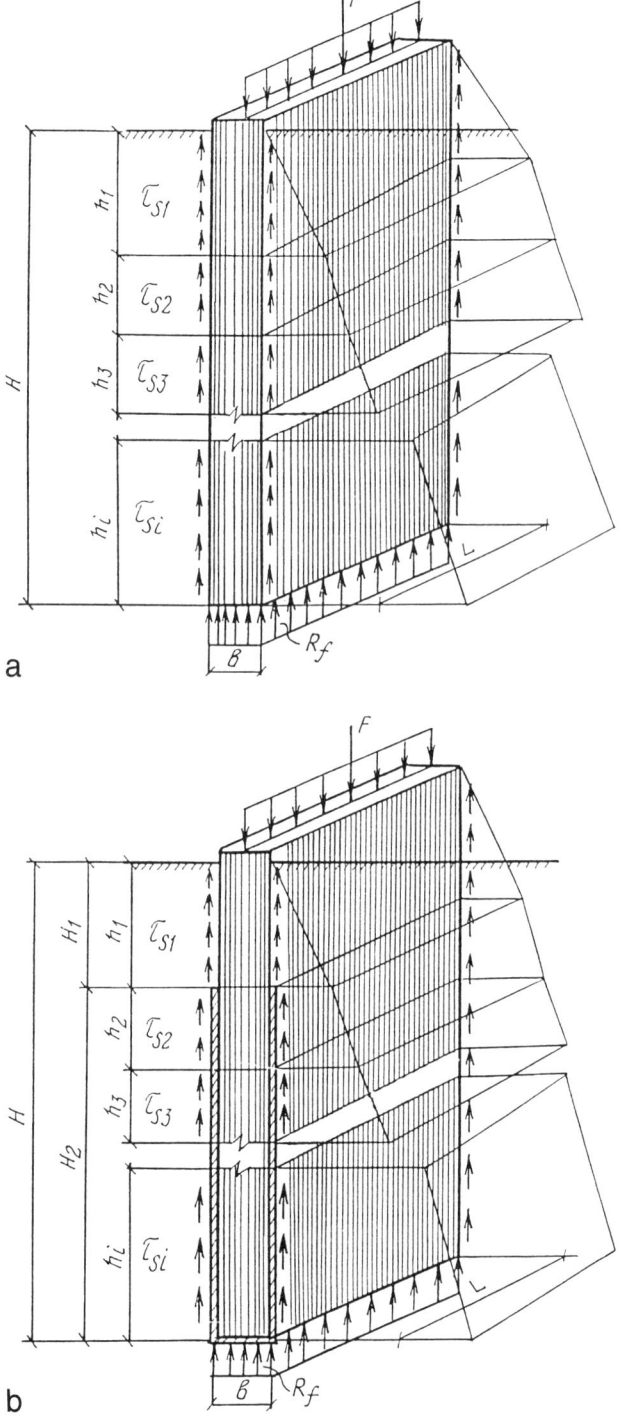

Figure 10.1. Design scheme of a trench foundation (rectangular in plan): a) diaphragm wall; b) with injected lower end and (partially) skin surface.

value of contact resistance to shear. In the conditions of constrained dilatancy it depends on the area of the sheared surfaces (see Section 2.2). A footing has four such surfaces: two skin surfaces, and two end surfaces. Consequently, the bearing capacity can be expressed as

$$F_s = 2F_{s1} + 2F_{s2} = 2(A_1 \bar{\tau}_{s1} + A_2 \bar{\tau}_{s2}) \tag{10.4}$$

where F_{s1}, F_{s2} are, respectively, the bearing capacity of one of the skin or end surfaces.

Taking into account (7.1)

$$F_s = 2l\Sigma(\lambda_{oi}\gamma_i h_i \operatorname{tg} \varphi'_{ci(1)} + \tau_{di(1)})h_i + 2b\Sigma(\lambda_{oi}\gamma_i h_i \operatorname{tg} \varphi'_{ci(2)} + \tau_{di(2)})h_i \tag{10.5}$$

Where $\varphi'_{ci(1)}$, $\tau_{di(1)}$ are, respectively, the angle of contact friction and the dilatant component of resistance to shear along the skin surface; $\varphi'_{ci(2)}$, $\tau_{di(2)}$ are, respectively, the same quantities for the end surface.

So, establishing parameters φ'_{ci} and τ_{di} either in a shear apparatus or by calculation according to (7.3), the values of coefficient of elastic resistance K (rigidity of elastic connection) should be set separately, i.e. according to area for skin surface and for end surface (see Section 2.4). If $A' < 0.20\,A$, then, to make calculation simpler, we suggest not to establish the value of τ_{s2} separately, but to add the area of end surfaces to the area of skin surfaces. Then

$$F_s = 2(L + b)\Sigma(\lambda_{oi}\gamma_i h_i \operatorname{tg} \varphi'_{ci} + \tau_{di(1)}) \cdot h_i \tag{10.6}$$

In the case when injection of skin surface is not done, it is recommended to establish the values of strength parameters as corresponding to the natural density of soil composition, taking into account the fact that a temporary support of the trench by clayey suspension restrains the elastic relief of the massif and limits soil violations in the contact layer (see Section 8.9). By way of example, mean resistances to contact shear in control-mix sands (see Section 4.1) are given in Table A2.8. The calculation was made according to expression (7.3) for sands with density of composition $I_D = 0.8$ and 0.6, i.e. for dense and medium-dense sands.

In the case of injection of part of the height of skin surface, strength parameters should be established as corresponding to dense composition of soil. Expression (10.4) in this case will take the form

$$F_s = 2(L + b)\Sigma(\lambda_{oi}\gamma_i h_i \operatorname{tg} \varphi'_{ci} + \tau_{di})h_{1i}$$

$$+ 2(L + b)\Sigma(\lambda_{oi}\gamma_i h_i \operatorname{tg} \varphi'_{ci} + \tau_{di}) \cdot (H_1 + h_{2i}) \tag{10.7}$$

where h_1, h_2 are, respectively, the heights of non-injected and injected parts of the skin surface (Fig. 10.1).

Examples of mean resistances to contact shear along the injected section of trench foundations calculated by this method are given for control-mix sands in Table A2.9. The height of the non-injected upper section is assumed to be equal to

170 *Load-holding capacity of a deep footing*

4 m. Contact friction along this section was calculated as for the case of free dilatancy. When $\bar{\tau}_s$ was calculated according to (7.3) for the injected section, the values of I_D were assumed to be equal to 1.0 and 0.8.

10.3 LOAD-HOLDING CAPACITY AT THE LOWER END

Resistance of soil to the pressing in of the footing develops in the conditions of internal bulge, the mechanism of which is analogous to that described for bore piles (Section 9.3). The adopted calculation scheme is represented in Fig. 10.2.

After the exhaustion of settlement needed for the formation of a wedge-shaped soil body under the sole AOA', further movement is possible at the expense of thrust of the two lateral quasi-elastic prismatic bodies AOB and A'OB'.

Let us consider the soil stress condition in the bulge area. In the initial stress condition, i.e. before the emergence of thrust of massif along the planes AB and $A'B'$, the pressure acting on these planes is determined by the sum of lateral pressure due to the soil's own weight, and additional pressure due to shear stresses along the contact surface σ_{ad}. In the conditions of limiting equilibrium the lateral thrust pressure σ_s reaches the value of passive pressure, i.e.

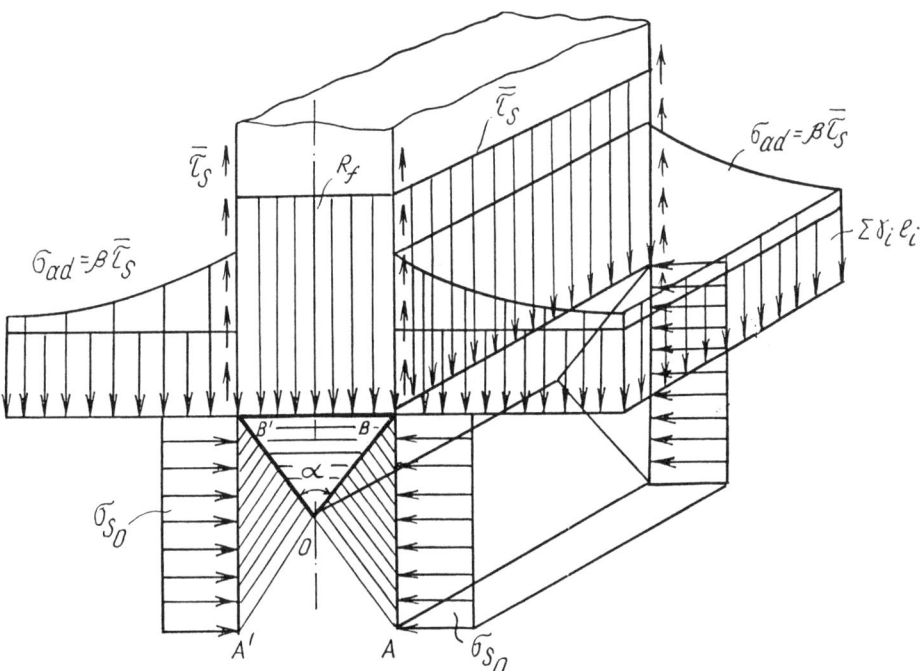

Figure 10.2. Design scheme for stresses in the internal bulge area under the lower end of a footing (rectangular in plan).

$$\sigma_s = \sigma_{so}\lambda_{pi}^2 + 2\tau_{di}\lambda_{pi} \tag{10.8}$$

Thrust pressure is distributed across the area of a prism with the side AB. If we express the height AB through the breadth of the footing sole as $AB = kb$, then the prism area is

$$A_s = kbL \tag{10.9}$$

where coefficient k corresponds to the relative depth of spreading of bulge area under the sole.

Vertical pressure R_f is distributed across the sole with the area $A_f = bL$. From the condition of equilibrium of forces

$$R_f A_f = \sigma_s A_s \tag{10.10}$$

where $R_f A_f = F_f$, i.e.

$$F_f = kbL\sigma_s \tag{10.11}$$

Then, having expressed σ_s according to (7.6), the bearing capacity at the lower end of a trench foundation can be written down as

$$F_f = kbL\left[\lambda_{oi}\lambda_{pi}^2\left(\Sigma\gamma_i h_i + \beta\bar{\tau}_s\right) + 2\lambda_{pi}\tau_{di}\right] \tag{10.12}$$

If we assume by analogy with piles that the internal bulge area spreads under the footing sole for two of its breadths, the $k = 2$. Correspondingly,

$$R_f \approx 2\sigma_s \tag{10.13}$$

It is recommended to establish additional pressure at the level of the sole σ_{ad} due to the shear stresses τ_s, for footings manufactured without injection of skin surface, proceeding from the mean value of $\bar{\tau}_s$ all along the height, while in the case when the skin surface is grouted, it is better to employ the mean value for the injected section. Resistances under the lower ends of trench foundations (injected and non-injected) in control-mix sands, calculated according to (7.12), are represented respectively in Tables A2.10 and A2.11.

10.4 EQUATION FOR CALCULATING THE TOTAL LOAD-HOLDING CAPACITY

Taking into account (10.4) and (10.11), we suggest the following calculation expression for the load-holding capacity of a trench foundation:

$$F = \gamma_c\Big\{\gamma_{cs}2(L + b)\bar{\Sigma}\left(\lambda_{oi}\gamma_i h_i \,\text{tg}\, \varphi'_{ci} + \tau_{di}\right)h_i$$
$$+ \gamma_{CR}nkbL\left[\lambda_{oi}\lambda_{pi}^2\left(\Sigma\gamma_i h_i + \beta\bar{\tau}_s\right) + 2\lambda_{pi}\tau_{di}\right] \tag{10.14}$$

where γ_c is the coefficient of work conditions of a trench foundation in soil; γ_{CR},

172 *Load-holding capacity of a deep footing*

γ_{cs} are the coefficients of soil work conditions respectively, under the sole and along the skin surface; n is the coefficient of accepted settlement which expresses the ratio of accepted settlement s_i to ultimate settlement s_{max} at which the maximum resistance to internal bulge is mobilized (see Section 9.2.2). When cement mortar is injected under the lower end, $n = 1$.

10.5 CONCLUSION

The same approach for working out suggestions for calculating the load-holding capacity of footings is used, as in the case of piles. In connection with the fact that this type of foundation has so far been studied rather fragmentarily, the possible directions for further research are more numerous than conclusions proper. At the same time the knowledge of peculiarities of non-cohesive soil behaviour in the conditions of constrained dilatancy can become the basis for refining and perfecting these constructions.

Refined strength conditions of dilating non-cohesive soils can be used as the basis for a method of calculation of trench foundations load-holding capacity in non-cohesive soils. Therefore, an introduction of adjustment for dilatancy allows for the calculating of contact resistance to shear taking into account the dimensions of skin surfaces of a trench foundation and to optimize these dimensions. The solution of the internal bulge calculation is the theoretical basis for a method of calculating resistance at the lower end of the footing.

Injection compression/compaction of the base can be a reserve for a considerable increase of the bearing capacity and effectiveness of trench foundations.

A number of directions for further research can be singled out.

1. Studying deformative-stress condition of soil around trench foundations of different configuration in plan, taking into account the technologies used for their manufacture.

2. Perfecting technologies with an aim to most fully mobilize the factor of constrained dilatancy as a component of load-holding capacity.

3. Studying the process of internal bulge in non-cohesive soil from under a rectangular loading place; solving in this connection the problems connected with additional stresses from contact friction along skin surfaces, and their experimental testing.

4. Accumulation, generalization and analysis of data of in-situ footing tests from the viewpoint of dilatancy and with an objective to establish justified coefficients of work conditions, reliability and acceptable settlement.

CHAPTER 11

Load-holding capacity of an injection anchor

11.1 CALCULATING SCHEME

In non-cohesive soils the load-holding capacity of an injection anchor is ensured by resistance to contact shear along the root surface; usually the form of the root in these cases is close to cylindrical. The degree of widening of the root relative to the initial hole diameter depends on the degree of compaction of soil in the contact zone (Sobolevsky, 1985).

Due to comparatively small cross section of the root, the diameter of which usually constitutes from 150 to 450 mm, a considerable influence is exerted on the mobilized values of τ_s by the character of distribution of stresses along the length of the body which has undergone injection. Anchors with a tension and compression root are distinguished (Fig. 11.1).

In the case of anchors with a tension root the transfer of load is effected by forces of fixing the tie with cement-stone body. The distribution of stresses along the length of the root is represented in Fig. 11.2, and Fig. 11.3 shows the dependence of the bearing capacity on the root length (Ostermayer & Sheele, 1977). An increase of tension in the tie causes tension stresses in the fixed cement of the root; these stresses can exceed the strength of the cement and cause cracking. The tension of the root is also accompanied by tension deformations in the contact layer, which cause a decrease of dilatancy during shear.

The distribution of stresses in an anchor with a compressed root (compression tube) is represented in Fig. 11.4 (Ivering, 1987). The cross thrust deformation accompanying the compression of the cement-stone body is a factor increasing dilatant stresses (Popov, 1989). So, while calculating contact resistance to shear, in the case of anchors we should also allow for the intrinsic deformations of the sheared body (Section 2.2).

The calculation scheme of an anchor with a tension and compression cylindrical root is represented in Fig. 11.5. The following geometrical parameters are distinguished: the length of the root L_r, the mean diameter of the root D_r, tendon diameter d_t, original hole diameter d_h, mean depth of the root embedment $\bar{d}_i$, the angle of inclination of the anchor to horizon Θ.

174 *Load-holding capacity of an injection anchor*

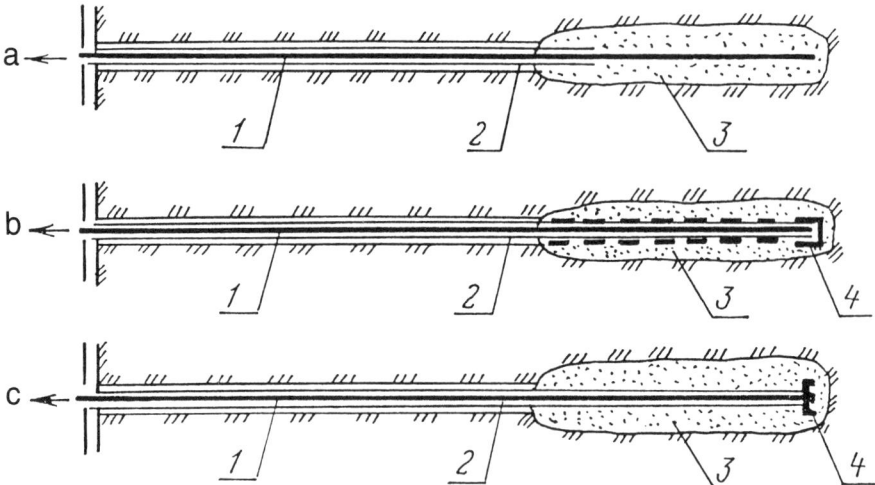

Figure 11.1. Types of anchor designs: a) with a tension root; b) with a compression root; c) with a base washer. 1. tendon; 2. protective sheath of tendon; 3. root; 4. compression tube or base washer.

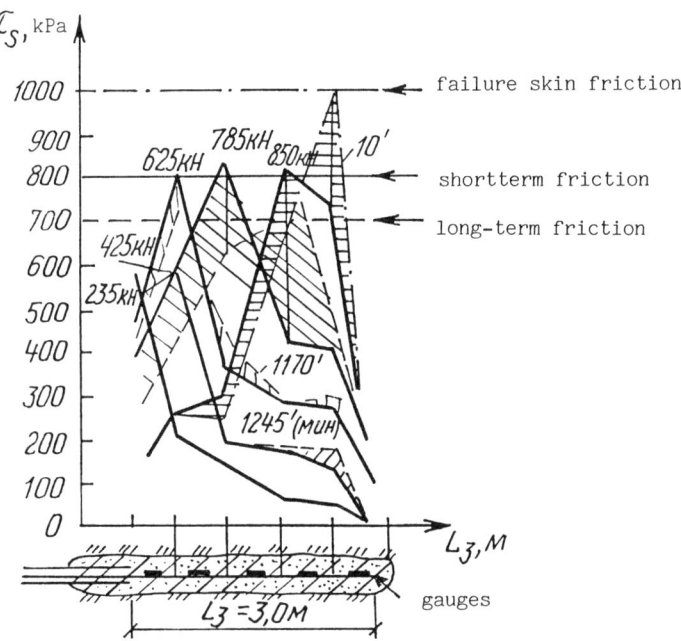

Figure 11.2. Distribution of stresses along the length of the root which is working for tension, according to Ostermayer & Scheel (1977). (L_b – the length of tendon embedment in the cement body).

Calculating scheme 175

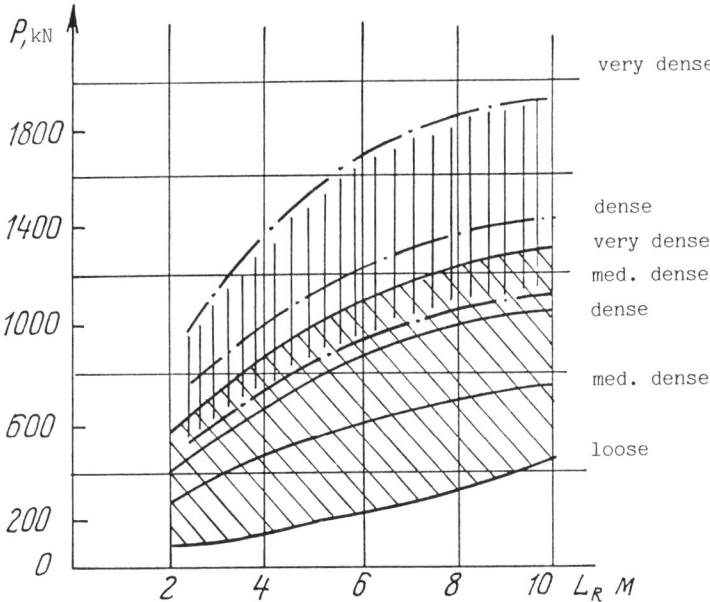

Figure 11.3. Functional relationship of the bearing capacity to the anchor root length in sands of different density (Ostermayer & Scheele, 1977).

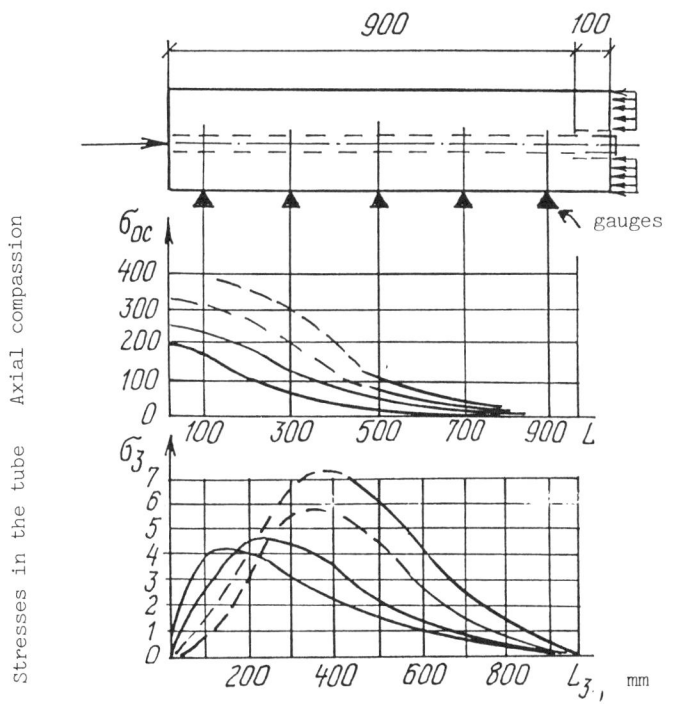

Figure 11.4. Distribution of stresses along the length of the anchor root (anchor with a compression tube) (Ivering, 1987).

176 *Load-holding capacity of an injection anchor*

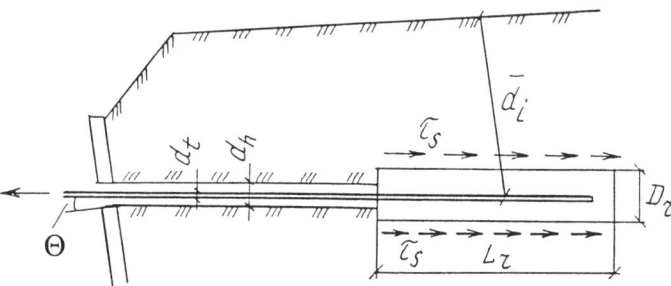

Figure 11.5. Calculation scheme for an anchor.

When the root is tensed its length increases by a certain value designated as Δl_r; during a compression there is a cross widening ΔD_r.

Before loading, the anchor root experiences normal pressure σ_{n_o} from the component of the natural soil weight on the level of the depth of embedment $\bar{d}_i$. The mean value of σ_{n_o} depending on the inclination of the anchor with respect to horizon can be defined as

$$\bar{\sigma}_{n_o} = \Sigma \gamma_i \bar{d}_i [\lambda_o + 0.5(1 - \lambda_o)\cos \Theta] \qquad (11.1)$$

where λ_o is the coefficient of lateral pressure of soil in quiescent state.

The ultimate load-holding capacity of an anchor in soil is calculated proceeding from the assumption that mean shear stresses $\bar{\tau}_s$ are evenly distributed along the root surface. It can be expressed as

$$P = \pi D_r L_r \bar{\tau}_s \qquad (11.2)$$

As for the value of $\bar{\tau}_s$, it depends on the intrinsic deformations of the root, i.e. on the anchor's design.

11.2 CONTACT RESISTANCE TO SHEAR ALONG THE ROOT SURFACE

If the cement body of the root did not undergo deformations during the loading of the anchor, the value of $\bar{\tau}_s$ could be calculated according to expression (7.6) or, using the symbols adopted in this work as

$$\bar{\tau}_s = \tau_u = a \sqrt[3]{\frac{2E}{(1 + \nu)D_r}} + \sigma_{n_o} \operatorname{tg} \varphi'_c \qquad (11.3)$$

But in reality the tension of the root accompanied by moving apart of contact layer grains causes a decrease of the dilatant thrust by a certain value - $\Delta \sigma_d$, with a parallel decrease of the contact friction with respect to the value expected according to (11.3). On the contrary, the compression of the root with its cross widening means an increase of the dilatant thrust by $+ \Delta \sigma_d$ (Hobst, 1987). These phenomena can be taken into account by introducing a special coefficient of root deformativity γ_{cr} (Popov, 1989). Then

$$\bar{\tau}_s = \gamma_{cr}\tau_u = \tau_u \pm \Delta\sigma_d \, \text{tg} \, \varphi'_c \qquad (11.4)$$

Below we shall deduce analytical functions for γ_{cr} according to this suggestion.

11.2.1 Resistance to shear with tension of the root

Fig. 2.3 represents a model reflecting the behaviour of non-cohesive soil during tension of the shear surface. The tension of the anchor root leads to moving apart of contact layer grains. This signifies a local decrease of density of their packing. Dilatancy, as was pointed out before, is practically proportional to the original soil density and is very sensitive to its decrease.

The moving apart of grains in the contact layer, occurring during the tension of the root, in conditions of constrained dilatancy will definitely lead to a decrease of stresses of dilatant thrust by a certain value – $\Delta\sigma_d$.

As can be seen in Fig. 11.2, the peak contact resistance to shear for tension-root anchors is mobilized at the head of the anchor and travels to its lower part after the unwedging influence of soil grains is overcome. The overcoming of resistance to shear in the upper section is compensated for by its increase at the remaining anchor length. Such compensation does not take place, if a grain 'flow' begins all along the root surface (Sobolevsky, 1985). In the condition, corresponding to the moment when this flow starts, the relative lengthwise deformation of the root with the in-built tendon will constitute

$$\varepsilon_{r/t} = \frac{\sigma_r}{E_{r/t}} = \frac{4D_r l_r \bar{\tau}_s}{E_r D_r^2 + E_t d_t^2} \qquad (11.4)$$

where E_r, E_t are, respectively, modula of elasticity of the root's fixed cement body and of the steel tendon.

Proceeding from the proposition regarding the hydrostatic effect of dilatant stresses (see Section 6.3), the relative deformation of the massif due to grain moving apart can be expressed as

$$\varepsilon_M = -\frac{\Delta\sigma_d}{E} \qquad (11.5)$$

Where the '–' signifies elastic relief of thrust stresses. From the condition of joint deformations of root shaft and soil in limiting equilibrium.

$$\varepsilon_{r/t} = \varepsilon_M \qquad (11.6)$$

consequently,

$$\Delta\sigma_d = \frac{4D_r L_r E \bar{\tau}_s}{E_r D_r^2 + E_t d_t^2} \qquad (11.7)$$

Substituting (11.7) to (11.4) we receive

178 *Load-holding capacity of an injection anchor*

Table 11.1. Values of γ_{cr}^{tens} for tension-root anchors.

Root length L_r, m	Tendon diameter d_t, mm	Root diameter D_r, m			
		0.15	0.20	0.25	0.30
4	25	0.54*	0.58	0.63	0.66
	40	0.61	0.63	0.66	0.68
	60	0.70	0.70	0.71	0.72
6	25	0.43	0.48	0.53	0.56
	40	0.51	0.53	0.56	0.59
	60	0.61	0.61	0.62	0.63
8	25	0.36	0.41	0.46	0.50
	40	0.44	0.46	0.49	0.52
	60	0.54	0.54	0.55	0.56

* Calculations were made with $E = 356$ MPa, $E_r = 2 \times 10^4$ MPa, $E_t = 2 \times 10^5$ MPa, $\varphi'_c = 30°$.

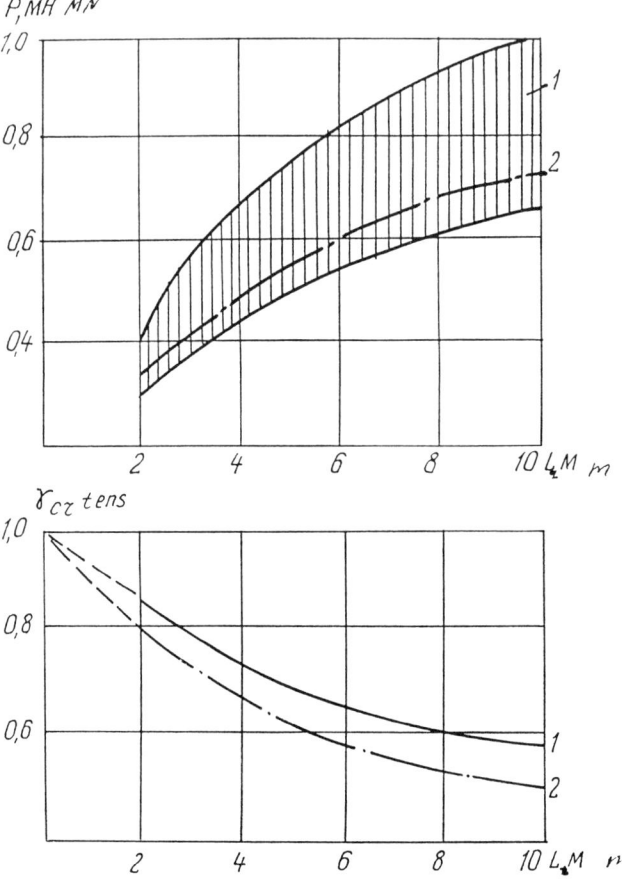

Figure 11.6. Functions $P = f(L_r)$ and $\gamma_{cr}^{tense} = f(L_r)$. 1. according to Ostermayer; 2. according to Equation (11.2) calculation.

$$\gamma_{cr}^{tens} = \frac{1}{1 + \dfrac{4D_r L_r E \operatorname{tg} \varphi'_c}{E_r D_r^2 + E_t d_t^2}} \qquad (11.8)$$

Coefficient γ_{cr}^{tens} takes into account the influence on the contact resistance to shear of the following values: root dimensions (L_r, D_r), modulus of elasticity of massif (E), modula of elasticity of the injected cement body (E_r) and steel tendon (E_t), the angle of contact friction of soil φ'_c. When the tendon diameter d_t is increased, for example if a steel tube is inserted along the length of the root (Fig. 11.1b) (Ivering, 1981; Hanna, 1982), the influence of root deformativity decreases and the value of γ_{cr}^{tens} approaches '1'.

The values of γ_{cr}^{tens} for tension-root anchors are represented in Table 11.1. The tendon diameter $d_t = 60$ mm corresponds to the case when a steel tube is inserted along the length of the root.

An analysis of Table 11.1 and function (11.8) reveals that an increase of root length L_r causes a decrease of the mobilized contact friction $\bar{\tau}_s$ and, correspondingly, of the relative bearing capacity of the anchor in soil. This proposition is substantiated by in-situ experiments conducted by Ostermayer & Sheele (1977). Fig. 11.6a, b represents comparative functions $P = f(L_r)$ and $\gamma_{cr}^{tens} = f(L_r)$ for anchors in gravelly medium-dense sand (Popov, 1989).

11.2.2 Resistance to shear during root compression

As the worldwide anchoring experience testifies, the efficiency of root embedment in rocky and non-cohesive soils considerably increases when the force is transferred to the root through its lower end. In such design an anti-corrosion casing envelops the tendon along its entire length, and the transfer of stresses in the embedment to the lower end of the root is ensured through a base washer or a compression tube. As Hobst (1987) points out, when anchor is loading the base washer rests in the hole fill, compresses it and causes radial stresses with a thrust and compression of the surrounding soil. Additional cross stresses in the root increase with the increase of the pressure transferred by the washer, depending on the value of Poisson coefficient of the root material. These thrust stresses are added to those appearing due to dilatancy of contact layer grains.

Let us establish the relative deformation of expanding of the root shaft. During the ultimate force application which preceeds the beginning of failure of the cement stone body, taking into account (11.2).

$$\varepsilon_r = \frac{\sigma_r \nu_r}{E_r} = \frac{4L_r \tau_s \nu_r}{E_r D_r} \qquad (11.9)$$

where ν is Poisson coefficient of the cement stone body, constituting approximately 0.20.

The radial elastic deformations of the massif will be expressed as

Table 11.2. Values of γ_{cr}^{comp} for compressed-root anchors.

Root length L_r, m	Root diameter D_r, m			
	0.15	0.20	0.25	0.30
3	1.20	1.14	1.11	1.09
4	1.28	1.20	1.15	1.11
6	1.49	1.33	1.25	1.19

* The calculations were made with $E = 356$ MPa, $E_r = 2 \times 10^4$ MPa, $\nu = 0.20$.

$$\varepsilon_M = \frac{\Delta\sigma_d}{E} \tag{11.10}$$

Proceeding from the conditions of joint relative deformations of the root and the soil in limiting equilibrium, equating (11.9) to (11.10), we receive

$$\Delta\sigma_d = \frac{4L_r \bar{\tau}_s \nu_r E}{E_r D_r} \tag{11.11}$$

Then, substituting (11.11) to (11.3), we get

$$\gamma_{cr}^{comp} = \frac{1}{1 - \dfrac{4L_r E \nu_r \, \text{tg}\, \varphi_c'}{E_r D_r}} \tag{11.12}$$

The values of γ_{cr}^{comp} for a number of characteristic root sizes, with the same assumed values as in the case of γ_{cr}^{tens} are given in Table 11.2.

Concentration of stresses near the base washer can cause the failure of the cement body of the root. But, as Hobst (1987) points out, this does not decrease the strength of the anchor tendon embedment in the soil, because the crushing of the fixed cement only leads to Poisson coefficient increasing up to 0.5.

11.3 CRITICAL LENGTH OF COMPRESSED ROOT

Failure of the cement body, even if it does not cause anchor failure in soil, is nevertheless dangerous in that it can cause the destruction of the tendon's anti-corrosion cover. This is unacceptable for permanent anchors and for anchors embedded in steel-aggressive soil mediums. Consequently, taking into account the danger of this type of destruction is a sine qua non for optimization of anchor design.

The fixed cement of the root works in the conditions of triaxial compression, which determines its strength as different from strength measured in standard sample compression tests. Sobolevsky (1985) proposed the principal of compressed root calculation on the basis of octahedral strength theory. Employing this principle with additional introduction of dilatant stresses, Popov (1989) suggested an expression for the critical length of compressed root L_r^{cr}, at which the aggregate mobilized resistance to shear does not exceed the strength of the cement body.

For analysis, let us isolate in the root body, an elementary unit in the most stressed section, and establish active principal stresses. According to the calculation scheme in Fig. 11.7, the least principal stress in the fixed cement is

$$\sigma_3 = \sigma_{rez} = \sqrt{\tau_s^2 + \sigma_s^2} = \frac{\bar{\tau}_s}{\sin \varphi_c'} = \frac{\gamma_{cr}^{comp} \tau_u}{\sin \varphi_c'} \tag{11.13}$$

Correspondingly, the highest principal stress is

$$\sigma_1 = \frac{P}{A_r} \cos \varphi_c' = \frac{4 L_r^{cr} \bar{\tau}_s}{D_r} \cos \varphi_c' = \frac{4 L_r^{cr} \gamma_{cr}^{comp} \tau_u}{D_r} \cos \varphi_c' \tag{11.14}$$

According to the octahedral theory, the limit of strength in a given point of solid medium is achieved, when the octahedral stress in this point reaches the value corresponding to the strength of the material with ordinary tension or compression (Filonenko-Borodich, 1961).

The tangent octahedral stress with $\sigma_3 = \sigma_2$ equals

$$\tau_{oct} = \frac{\sqrt[3]{2}}{3}(\sigma_1 - \sigma_3) \tag{11.15}$$

In the limiting state during ordinary compression $\sigma_3 = \sigma_2 = 0$, and

$$\sigma_1 = R_r^{comp} \tag{11.16}$$

where R_r^{comp} is the strength of the cement-stone body with respect to compression. Then

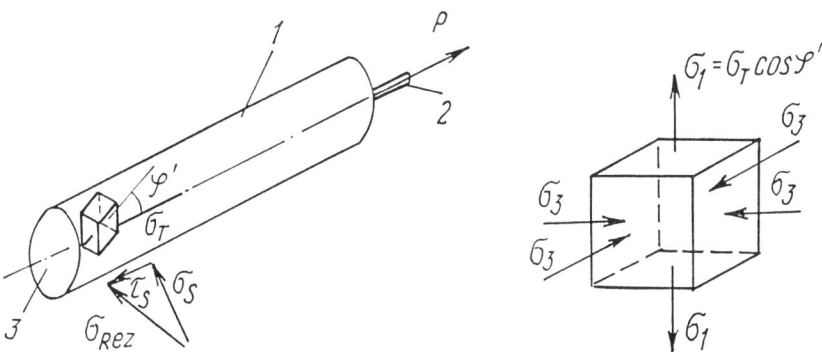

Figure 11.7. Calculation scheme for establishing the strength of a compressed root: a) stress condition of the root; b) stress condition of the elementary volume. 1. root; 2. tendon; 3. base washer.

$$\tau_{oct} = \frac{\sqrt{2}}{3} R_r^{comp} \tag{11.17}$$

The condition of strength can then be written down as

$$\sigma_1 - \sigma_3 \leq R_r^{comp} \tag{11.18}$$

The values of R_r^{comp} can be established by tests according to norms for concrete testing, for instance GOST 10180-78. In this case it is necessary to take into account the final water-cement ratio in the cement mortar, which is established in the hole consequent to the injection (Mishakov, 1980; Sobolevsky, 1986).

Substituting (11.13) and (11.14) to (11.18) we receive

$$L_r^{cr} = \frac{D_r}{4 \cos \varphi_c'} \left[\frac{R_r^{comp}}{\gamma_{cr}^{comp} \left(a \sqrt[3]{\frac{2E}{(1+\nu)D_r}} + \sigma_{n_o} \mathrm{tg}\, \varphi_c' \right)} + \frac{1}{\sin \varphi_c'} \right] \tag{11.19}$$

11.4 EQUATION FOR CALCULATING THE LOAD-HOLDING CAPACITY

The load-holding capacity of an injection anchor can be calculated as

$$P = \frac{\gamma_{cr}\gamma_{cs}}{\gamma_s} \pi D_r l_r \left[a \sqrt[3]{\frac{2E}{(1+\nu)D_r}} + \Sigma \gamma_i d_i \{\lambda_o + 0.5(1-\lambda_o)\cos\Theta\} \mathrm{tg}\, \varphi_c' \right] \tag{11.20}$$

where γ_{cs} is the coefficient of work conditions of the anchor during the 'root-soil' contact; γ_s is the coefficient of the reliability for 'root-soil' contact.

The diameter of the injected root body D_r can be calculated according to (9.22).

A correct establishing of work conditions coefficient, and especially of coefficient of reliability, when calculating the work load for the anchor, is very important. Unlike in the case of piles, where usually only additional settlements occur when the acceptable load is exceeded, extra load of an anchor can result in a spontaneous failure of the whole construction.

An analysis of coefficients γ_{cs} and γ_s is given in Sobolevsky (1985). Coefficient γ_{cs} is introduced in order to allow for possible changes of soil stress condition around the root. These changes can be caused by a rising of the underground water level higher than the root embedment level, by taking a layer of soil from the surface etc. When designing anchors, especially permanent ones or those which are part of important constructions, it is necessary to allow for the influence on the soil condition of underground sewage and water-supply systems, because leakages and breakages of those systems can exert a negative influence on the

Table 11.3. Recommeded reliability coefficients γ_s for anchor calculation.

Degree of failure risk		Temporary anchors		Permanent anchors	
Category	a) casualty hazards b) material damage	Anchor class	γ_s	Anchor class	γ_s
Low	a) absent (the construction area is not populated) b) minimal	1	1.6	4	1.8
Considerable	a) not great (the construction area is scarcely populated) b) considerable consequences for surrounding constructions	2	1.6	5	2.0
High	a) great (the construction area is densely populated) b) serious consequences for vital constructions	3	1.8	6	2.0

work of the anchor. If such unfavourable factors are present, it is recommended to assume γ_{cs} to be equal to 0.8, while if they are absent $\gamma_{cs} = 0.9$.

We should also differentiate between reliability coefficients for single anchors and anchors that are a part of a system. The breakage of one or several anchors in tie-back wall does not necessarily mean that the whole structure undergoes failure (Stelle, 1976), whereas the exhaustion of the bearing capacity of the anchor supporting suspension system is fraught with catastrophic consequences.

An attempt to take into account these peculiarities is realized in Hong Kong norms (1980), where on the basis of standard requirements of different countries a table of γ_s values has been worked out, allowing for potential failure dangers. We suggest making use of this table in this work (see Table 11.3).

11.5 GROUP EFFECT OF ANCHORS

The mutual influence of anchors in a construction can be determined by
– injection conditions during formation of the root;
– redistribution of stresses between anchors depending on the rigidity of the elements of the supported construction;
– superimposition of stresses appearing in the areas adjacent to neighbouring anchor roots.

Injection conditions, the phenomena of hydrorupture, filtration compaction, contact filtration have been dealt with Sobolevsky (1984, 1985, 1986). The phenomenon of redistribution of stresses in the constructions of retaining walls support was described by Stille (1976). Below we consider the mutual influence resulting from the superimposition of stresses around the roots of adjacent anchors, taking into account the additional stresses caused by dilatancy.

The stresses of dilatant thrust appearing after the anchor is loaded are distributed through a certain round area around the root. We shall designate this area's radius as the root influence zone/radius (Sobolevsky, 1985). When the distance between roots is not large enough, their zones of influence can cross. The aggregate stresses resulting from such crossing are capable of causing additional soil deformations between the roots and, consequently, a decrease of the bearing capacity of the adjacent anchors (Fig. 11.8).

Let us designate the root influence radius as R_r. In order to calculate it, we shall employ an axis-symmetrical sum about a hollow thick-walled discus which experiences evenly distributed internal and external pressure (Bezukhov, 1968; Sobolevsky, 1985). The value of internal pressure P_a corresponds to the sum of the values of dilatant thrust and natural pressure regenerated at the root contour by injection, i.e.

$$P_a = (\sigma_d + \sigma_{n_o})\gamma_{cr} \qquad (11.21)$$

while the value of external pressure corresponds to the natural pressure of soil (11.1) or

$$P_b = \sigma_{n_o} \qquad (11.22)$$

Tangential stresses in the wall of the cylinder under study are expressed as follows (Bezukhov, 1968);

$$\sigma_\Theta = \frac{P_a r_r^2 - P_b R_r^2}{R_r + r_r} + \frac{(P_a - P_b)r_r^2 R_r^2}{(R_r^2 - r_r^2)r_i} \qquad (11.23)$$

If we assume that at the external contour of the root influence zone $r_i = R_r$, $\sigma_\Theta = \sigma_{n_o}$ then

$$R_r = \frac{D_r}{2}\sqrt{\frac{\gamma_{cr}(\sigma_{n_o} + \sigma_d)}{\sigma_{n_o}}} =$$

$$= \frac{D_r}{2}\sqrt{\gamma_{cr}\frac{a/\text{tg}\,\varphi_c'\sqrt[3]{\dfrac{2E}{(1+\nu)D_r}} + \sigma_{n_o}\text{tg}\,\varphi_c'}{\sigma_{n_o}\text{tg}\,\varphi_c'}} \qquad (11.24)$$

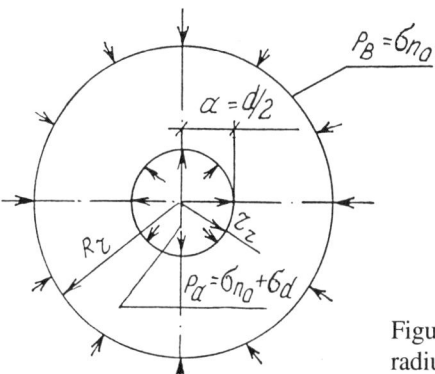

Figure 11.8. Calculation scheme for establishing the radius of root influence zone after the anchor is loaded (contour dilatancy).

It follows from equation (11.24) that the value of R_r depends on the root diameter, physic-mechanical soil characteristics, initial pressure and the method of tendon embedment (γ_{cr}).

As calculation reveals, in order to exclude the mutual influence of anchor roots due to the superimposition of dilatant stresses, the minimum distance between them should be assumed to be

$$B_{min} \nless 2R_r \tag{11.25}$$

This happens in all circumstances on condition that

$$B_{min} \geq 4\ D_r \tag{11.26}$$

11.6 CONCLUSION

The suggested calculation method of a single injection anchor was experimentally tested during construction of the Minsk metro, and it appears possible to recommend it for practical employment in the suggested form.

Resistance to contact shear developing in non-cohesive soil compacted by injection in conditions of constrained dilatancy is the principal source of the bearing capacity of an injection anchor. Mean values of contact resistance to shear depend on the lengthwise and cross deformations of the anchor root during the process of loading.

The tension of the root under loading decreases the mobilized contact resistance to shear along the root length due to a decrease of the value of dilatant thrust. On the contrary the compression of the root causes an increase of the dilatant thrust and, consequently, of mean contact resistance to shear.

Taking into account additional stresses from dilatancy and their influence on the stress condition, it is possible to evaluate the strength of the material of the anchor root. It means that the suggested method opens up possibilities for optimization of the length, diameter and material of anchor root and its steel tendon.

As a direction for further research a study of anchors working in a system, and a refinement of coefficients of work conditions and reliability would be recommended.

CHAPTER 12

Sketches of several dilatancy manifestations

12.1 REINFORCED EARTH AS A COMPOSITE MATERIAL

The essence of soil reinforcement is its strengthening in the directions of potential shears. For this purpose various materials, receiving dilatant stresses developing in the soil, are introduced into the massif. Joint work of the reinforcing element and the soil is ensured by the forces of contact friction.

Figs 12.1-12.3 represent several cases of employing reinforcement for strengthening bases and supporting slopes. Reinforcing the base of a reservoir in Fig. 12.1 ensures the transfer of pressure to a larger area of the embankment situated on weak soil. Introduction of reinforcement into the high wall of a steep mountain slope in Fig. 12.2 reduces the possibility of landslides. Finally, positioning of reinforcing elements in a road bank ensures construction of an embankment in constrained conditions with steepness exceeding the angle of natural sloping.

Several principal schemes of soil reinforcement are employed in engineering practice. They are continuous reinforcement (soil is covered all throughout the reinforced area with the reinforcing material interspersed with backfill compacted layer by layer) and discrete reinforcement (the reinforcing material is positioned in the massif by separate elements).

For reinforcing the non-violable soil the so-called nailing is used: The constructing slope or excavation is fixed step-by-step with steel rods grouted into holes with a diameter of 50-80 mm. Sometimes rods with a diameter of 20-40 mm are just vibro-immersed into the massif without grouting. Due to similarity of technologies we can classify as reinforcement methods geomembranes and 'pile walls' (injection pile 'palisades', Fig. 12.4).

During artificial structure erection sandy and coarse grained soils are used, as a rule; these soils ensure high values of contact friction along the surface of the reinforcing elements. The use of soils containing clayey and silty fractions is limited due to the changeability of their strength parameters during moistening.

A wide spectrum of various materials with a large range of deformative properties is used for reinforcement, including filtrating and non-filtrating geotextiles, films, nets, gabions manufactured from wire, dense polyethylene, polyester,

Reinforced earth as a composite material 187

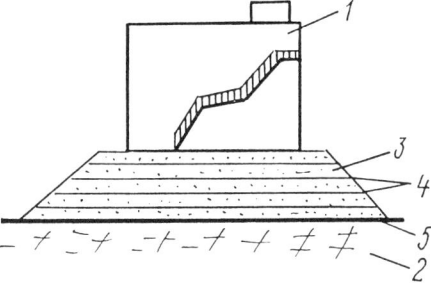

Figure 12.1. Reinforcement of a reservoir base: 1. reservoir; 2. weak soil; 3. embankment; 4. geotextile reinforcing layers; 5. geomembrane (non-filtering geotextile or film).

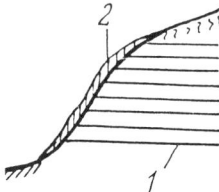

Figure 12.2. Reinforcement of a steep mountain slope by nailing: 1. nails (reinforcing rods grouted in holes or driven into soil); 2. shotcreting of reinforcement net along the slope surface.

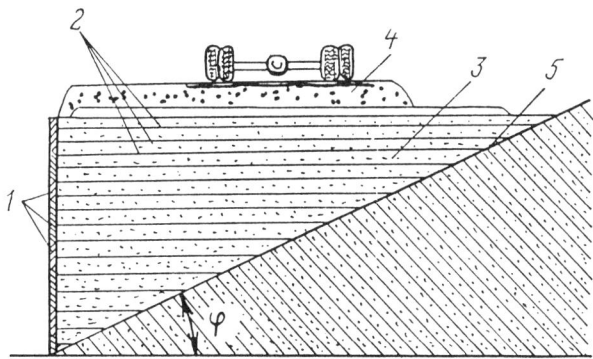

Figure 12.3. 'Classic' construction of a soil-reinforced structure: 1. facing; 2. reinforcement elements; 3. backfill layers; 4. road surface; 5. natural slope.

kapron, nylon. Galvanized (zinc-plated) or anodized steel bands, more rarely plastic or glassfiber fabric bands, are used for discrete reinforcement.

The bearing capacity of reinforced soil is ensured provided two conditions are met, to wit:

1. Shear stresses at the contact surfaces of the reinforcing elements should not exceed their ultimate value i.e.

$$\tau_i < \tau_u \tag{12.1}$$

2. Development of shear stresses should not cause rupture of the reinforcing elements themselves. i.e.

188 *Sketches of several dilatancy manifestations*

$$T_i < T_u$$

where T_i, T_u, are, respectively, mobilized and ultimate stretching forces.

There is a third additional condition: Plastic tension of the reinforcing element material should not exceed the development of acceptable plastic deformations of soil

$$\varepsilon_{pl,r} < \varepsilon_s \qquad (12.3)$$

This condition is often not stipulated.

When $\varepsilon_{pl,r} \geq \varepsilon_s$ the reinforcement material will follow the soil movements without offering any resistance to them. Moreover, the mechanical anisothropy with such ratio of deformation properties will be characterized by the weakening of soil in the directions of the material introduced therein; each layer of the introduced material becomes then a analogue of a potential surface of sliding (rupture).

The principal role in ensuring joint work of the reinforcing elements and soil belongs to contact friction. Let us analyze the conditions of mobilization of contact friction with continuous and discrete reinforcement, taking account of the dilatancy factor.

Continuous reinforcement corresponds to the case of free dilatancy. Mobilization of contact friction on the surfaces of films, nets, geomembranes installed over

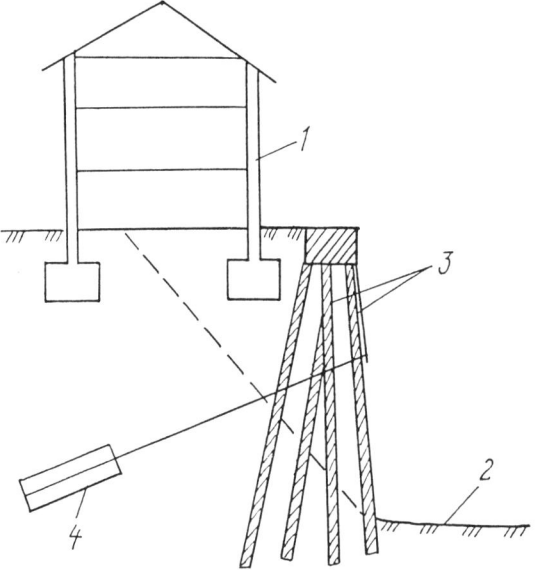

Figure 12.4. Supporting 'bar wall' consisting of an injection pile palisade: 1. existing structure; 2. excavation in soil; 3. injection piles built at intervals of 0.5 to 1.5 m; 4. soil anchors.

Reinforced earth as a composite material 189

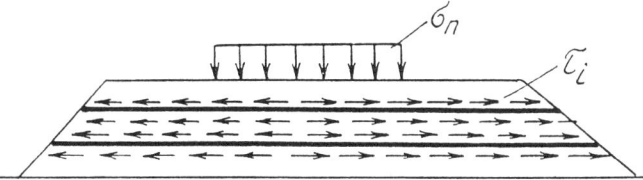

Figure 12.5. Transformation of external pressure into shearing forces along contact surfaces of reinforcing layers laid out over a large area.

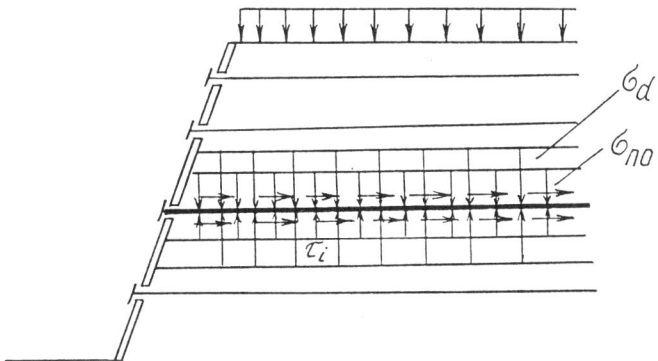

Figure 12.6. Stress condition of a 'narrow' reinforcing band or a nail jammed in the soil massif.

large areas is not accompanied by the emergence of dilatancy thrust stresses (Fig. 12.5). The dilatancy of contact shear occurring in the process of shear signifies an 'uplifting' of the backfill soil lying uppermost (see Section 1.6). In this case the ultimate resistance to shear τ_u is determined by normal pressure consisting of the soil's inherent weight σ_{n_o} and external loading along the soil surface, and by the value of the angle of contact friction with free dilatancy φ_o, i.e.

$$\tau_u = \sigma_{n_o} \operatorname{tg} \varphi_o \qquad (12.1)$$

In addition, it is absolutely necessary to take into account the influence on the parameter φ_o of contact loosening of the backfill due to relative stretching of the reinforcing elements. Evidently, φ_o will correspond not to the original density of the backfill, but to the density which is achieved at the moment when contact resistance shear is overcome.

In the case of plastic tension of the reinforcement element the soil composition in the contact layer can change from dense to that corresponding to critical density or even less. At the moment when critical density is achieved, conditions for the sliding of the reinforcing elements are created, joint work of the element and the soil is frustrated; but similar phenomena are, as a rule, realized on a limited section of the reinforcing elements length, where the larger shearing stresses are

190 *Sketches of several dilatancy manifestations*

concentrated. The reinforced soil in this place ceases to be a composite material.

It is rather difficult to allow for these phenomena while interpreting soil tests in shear apparatus, which is why a special scheme of contact shear apparatus of reinforcing elements DACS-A (see Section 3.5) is suggested, where an attempt was made to model joint work of soil and the material reinforcing it.

Now let us consider the conditions of shear of a separate reinforcing band and nail, which are jammed into the soil massif (Fig. 12.6a, b). In accordance with the concepts set forth in Chapters 2, 4 and 7, when shear takes place in soil with density exceeding the critical value, we can observe the development of dilatant thrust stresses which are superimposed on the initial normal pressure of the weight of the backfill soil σ_{n_o}. In the absence of a band stretching the ultimate contact resistance to shear in non-cohesive soil will be expressed, according to (7.1), as

$$\tau_u = \tau_d + \sigma_{n_o} \operatorname{tg} \varphi'_c = (\sigma_d + \sigma_{n_o}) \operatorname{tg} \varphi'_c \qquad (12.2)$$

where σ_{n_o} is the normal pressure of inherent weight and of the load on the soil surface.

A relative lengthening of the reinforcing element leads to a lower soil density with the decrease of dilatant thrust. It can be expressed as

$$\tau_u = (\sigma_d - \Delta\sigma'_d + \sigma_{n_o}) \operatorname{tg} \varphi'_c \qquad (12.3)$$

where $-\Delta\sigma'_d$ is the decrease of dilatant thrust due to the linear tension of the contact surface (see Section 2.2).

In the elastic stage of the reinforcement material's work its relative deformation is

$$\Delta\varepsilon_{el,r} = \frac{\Delta\sigma_r}{E_r} \qquad (12.4)$$

where $\Delta\sigma_r$ is the stretching stress; E_r is the modulus of elasticity of the reinforcement material.

The relative deformation of the massif occurring during the taking apart of grains can be expressed as

$$\varepsilon_M = \frac{\Delta\sigma'_d}{E} \qquad (12.5)$$

where E is the modulus of elasticity of soil.

From the condition of joint deformations in the elastic stage and from the concept set forth in Chapter 6 regarding the hydrostatic action of dilatant stresses, equalling (12.4) to (12.5) we receive

$$\Delta\sigma'_d = -\Delta\sigma_r \frac{E}{E_r} = -\Delta\sigma_r k_r \qquad (12.6)$$

The stretching stress in a flat reinforcing element can be expressed as

$$\Delta\sigma_r = \frac{\Delta T_{r,i}}{b_r t_r} \tag{12.7}$$

where $\Delta T_{i,r}$ is the stretching force; b_r, t_r are, respectively, the breadth and thickness of the reinforcing element.

For a nail, i.e. reinforcing element with round cross section,

$$\Delta\sigma_r = \frac{4\Delta T_{r,i}}{\pi d_N^2} \tag{12.8}$$

where d_N is the diameter of nail.

Substituting (12.7) and (12.8) into (12.6) we get,

$$\Delta\sigma_d' = -\frac{\Delta T_{r,i}}{b_r t_r}\frac{E}{E_r} \tag{12.9}$$

and

$$\Delta\sigma_d' = -\frac{4\Delta T_{r,i}}{\pi d_N^2}\frac{E}{E_r} \tag{12.10}$$

Coming back to (12.3) let us draw the readers' attention to the fact that when $\sigma_d = \Delta\sigma_d'$ dilatancy completely disappears, and when $\sigma_d < \Delta\sigma_d'$ it turns into contraction. The condition of constrained dilatancy is met only when $\sigma_d > \Delta\sigma_d'$. Other things being equal, mobilized contact resistance to shear is in reverse proportion to the ratio $k_r = E/E_r$ characterizing mechanical anisothropy of a composite material.

Let us write down (12.3) allowing for (12.9), (12.10) and the empirical expressions for τ_d (4.6), (4.7) and for φ_c' (4.9), (4.10). In the case of a reinforcing band

$$\tau_u = a\sqrt[3]{\frac{E}{(1-v^2)\omega_r b_r}}$$

$$-\left(\frac{T_r E}{2b_r t_r E_r} + \sigma_{n_o}\right)\mathrm{tg}\left[\varphi_\mu + (\varphi_o - \varphi_\mu)\exp-\alpha\frac{E}{(1-v^2)w_r b_r}\right] \tag{12.11}$$

In the case of reinforcing with nails

$$\tau_u = a\sqrt[3]{\frac{E}{(1+v)d_N}}$$

$$-\left(\frac{4T_r E}{\pi d_N^2 E_r} + \sigma_{n_o}\right)\mathrm{tg}\left[\varphi_\mu + (\varphi_o - \varphi_\mu)\exp-\alpha\frac{2E}{(1+v)d_N}\right] \tag{12.12}$$

Modulus of elasticity of a reinforced cement nail E_r constitutes about 2×10^4 MPa, i.e. with $E = 200\text{-}400$ MPa the ratio k_r is within the range 0.01-0.02.

The analysis of equations (12.11) and (12.12) allows to assess the influence on

192 *Sketches of several dilatancy manifestations*

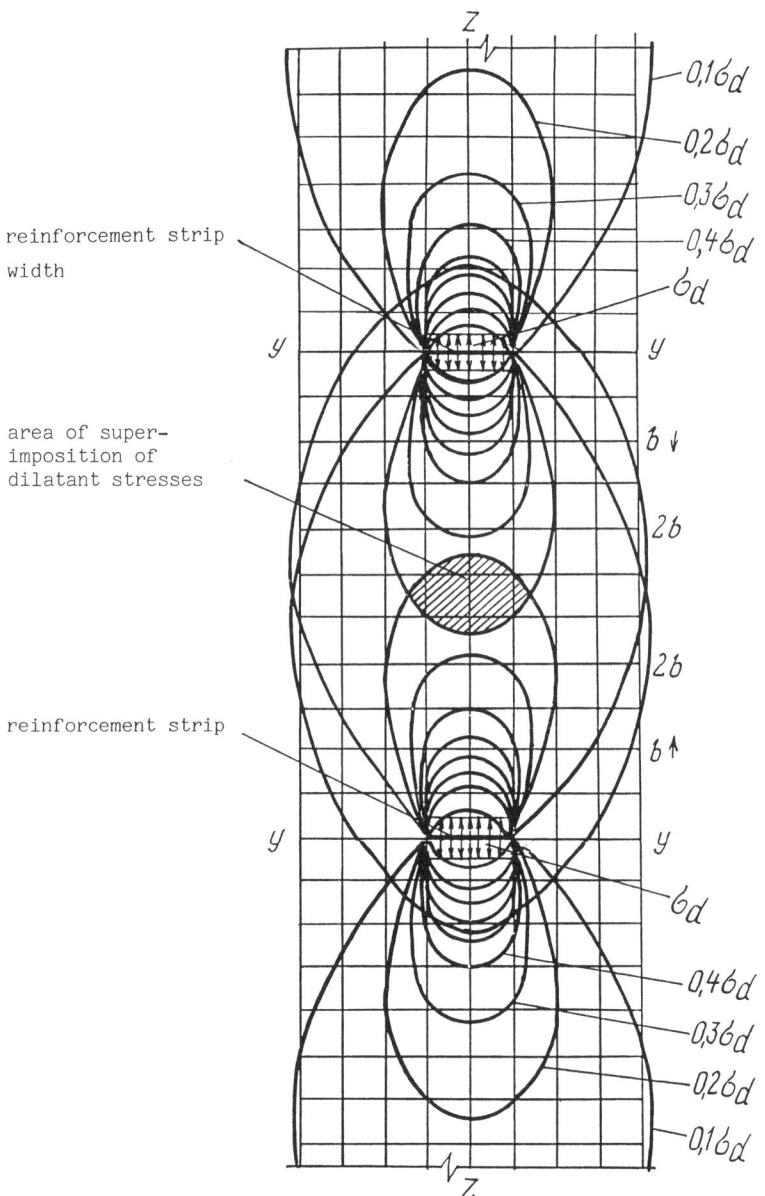

Figure 12.7. Intersection of zones of additional compressing stresses caused by dilatancy (isobars) for a linearly deformed semispace. Area of superimposition of dilatant stresses.

resistance to shear of various combinations of soil parameters, material parameters, geometrical characteristics of the reinforcing elements, which opens up possibilities for optimization of reinforced soil as a composite material. Table A2.12 provides an example illustrating the influence of grain size and soil density, geometrical parameters of the reinforcing elements (bands) and ratio k_r on the contact resistance to shear.

Table 12.1. Values of vertical additional stresses $\Delta\sigma_{d(z)}$ in shares of intensiveness of dilatant thrust $\Delta\sigma_d$ for flat reinforcing elements.

Z/b	Values of y/b					
	0	0.25	0.5	1.0	1.5	2.0
0	1.00	1.00	0.50	0.00	0.00	0.00
0.25	0.96	0.90	0.50	0.02	0.00	0.00
0.50	0.82	0.74	0.48	0.08	0.02	0.00
0.75	0.67	0.61	0.45	0.15	0.04	0.02
1.00	0.55	0.51	0.41	0.19	0.07	0.03
1.25	0.46	0.44	0.37	0.20	0.10	0.04
1.50	0.40	0.38	0.33	0.21	0.11	0.06
1.75	0.35	0.34	0.30	0.21	0.13	0.07
2.00	0.31	0.31	0.28	0.20	0.14	0.08
3.00	0.21	0.21	0.20	0.17	0.13	0.10
4.00	0.16	0.16	0.15	0.14	0.12	0.10
5.00	0.13	0.13	0.12	0.12	0.11	0.09
6.00	0.11	0.10	0.10	0.10	0.10	0.00

In the case of discrete reinforcement, when the condition $\sigma_d > \Delta\sigma'_d$ is met, dilatant thrust stresses appear in the process of development of contact shear on the surface of reinforcing elements. These stresses are distributed in the surrounding soil, creating fields of local additional stresses.

To clarify this we shall use a classic solution for soil stresses in the case of a flat sum (Tsytovich, 1963). Let us represent dilatant thrust as normal evenly distributed load. Fig. 12.7 shows the distribution of equal compressing stresses $\Delta\sigma_{d(z)}$ (isobars) in a linearly deformed massif. Dilatant thrust plays the role of the massif-compressing stresses, which is why isobars are plotted on both sides of the plane.

If we place two reinforcing elements vertically, isobars will intersect and the local concentration of stresses in the massif will be expressed in addition to dilatant pressures. Additional compression appears in the soil between the reinforcing elements. If we remember that the values of dilatant stresses can constitute up to 100, 150 and more kPa, it appears that such compression is worthy of special treatment and consideration.

The values of stresses $\Delta\sigma_{d(z)}$ in shares of intensiveness of the evenly distributed load $\Delta\sigma_d$ in the case of a flat shearing surface are represented in Table 12.1. It is clear from this table that compressing stresses achieve, for, example, about 50% of $\Delta\sigma_d$ at the distance of $1.12z/b$ along the band's axis of symmetry, 30% at the distance $2z/b_r$ and 10% at the distance $6z/b_r$. When these stresses are active in the soil between two bands, the stresses are added up.

When a natural slope is reinforced with nails, additional stresses in the massif caused by dilatancy decrease in the reverse proportion to the distance squared

194 *Sketches of several dilatancy manifestations*

(Fig. 12.8). The values of additional stresses $\Delta\sigma_{d(R)}$ in shares of intensiveness of dilatant thrust $\Delta\sigma_d$ for reinforcing elements with round cross sections are represented in Table 12.2.

The appearance of additional stresses in the reinforced massif due to contact dilatancy is one of the factors of stability. In the absence of conditions for lateral expanding of soil between reinforcing elements the superimposition of dilatant stresses on the stresses active in the soil causes an increase of stress at grain contact points.

Placing reinforcing elements at certain distances it is possible to ensure the development in the massif a field of close or equal local stresses. For example, when the vertical spacing between bands is equal to $6b_r$, for example, the minimal stresses along the axis of symmetry will constitute $\sigma_{n_o} + 0.4\Delta\sigma_d$. In this way it is possible, while designing reinforced soil, allowing for the stresses acting therein and varying the position and even breadth of the reinforcing elements, to ensure conditions of equal strength of the massif.

For instance, when the band's breadth is $b_r = 0.15$ m, the condition $z = 6b_r$ corresponds to the distance between the reinforcing elements equal to 0.9 m. Let it be noted that reinforcing element layers are usually designed to be placed at a distance of 0.5-1.2 m from one another.

When reinforcement is effected by means of nails a more intensive dispersion of stresses within the massif takes place (see Table 12.2). Therefore, the distance $6d_N$ will correspond to the minimal stress $\sigma_{n_o} + 0.006\Delta\sigma_d$. Nails are usually

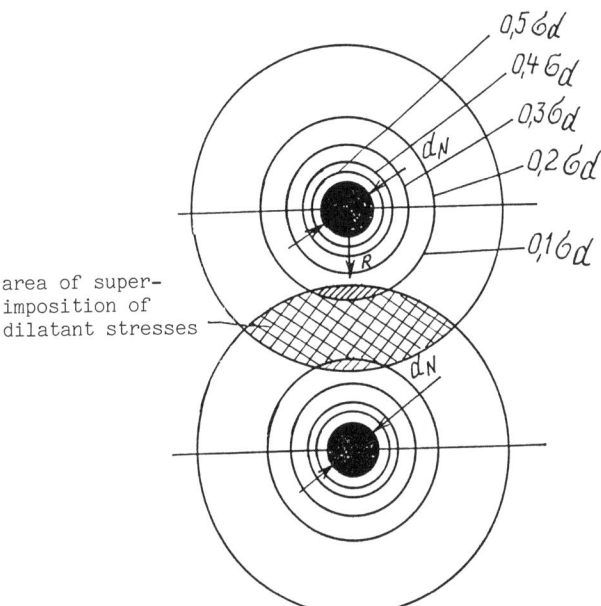

Figure 12.8. Intersection of zones of additional compressing stresses caused by dilatancy for round reinforcing elements (nails). Area of superimposition of dilatant stresses.

Table 12.2. Values of radial additional stresses $\Delta\sigma_{d(R)}$ in shares of intensiveness of dilatant thrust $\Delta\sigma_d$ for reinforcing elements with round cross sections.

R/d_n	0.5	0.75	1.0	1.5	2.0	2.5	3.0	4.0	5.0	6.0
$\sigma_d/\Delta\sigma_{d(R)}$	1.00	0.44	0.25	0.11	0.06	0.04	0.03	0.02	0.01	0.006

designed to be placed at the distance of 0.5-1.0 m from one another, more rarely over 1.2 m. When the diameter of a nail is about 80 mm this corresponds to 6-12 d_N.

The contents of this section are essentially a formulation of objectives for a study of reinforced soil as a composite material. Up to now the approaches to its research and calculation have been analogous to those employed for the majority of foundations, i.e. the stress condition of a separate reinforcing element is considered, shear stresses on its surfaces are established, the backfill material is chosen etc. In the suggested approach a combination of strength and deformative characteristics of soil and the reinforcing elements in it, with their mutual influence, is the key element.

Research of reinforced soil behaviour with different combinations of loads and reinforcement patterns can be conducted in large-size triaxial apparatuses (with sample diameters 200 mm and more). Tests based on the free dilatancy method will enable the researchers to model the case of continuous reinforcement, while constrained dilatancy tests will give a picture of discrete reinforcement (see Section 3.6). A series of similar tests in a triaxial apparatus with samples with a diameter up 203 mm and a height of up to 390 mm are described by Christian et al. (1986).

12.2 DISTORTION AND LIQUEFACTION OF SANDS

A number of accidents due to a sudden liquefaction of water-saturated sandy slopes and bases have been described in the practice of hydrotechnical and hedromeliorative construction (Gersevanov & Polshin, 1948; Ivanov, 1985; Maslov, 1982). Very often the reasons remain undisclosed, but in all the cases it is noted that liquefaction affected loose saturated soils.

In 1935 Kadomsky performed a simple experiment on the transfer of water-saturated sand into a liquefied state. Dry sand was poured in a thin stream into a water-filled vessel fitted with glass piezometers. Upon settlement the sand formed a loose water-saturated mass, on the surface of which a weight was installed. Then the sand mass was pierced by an ordinary ruler, which led to the failure of the loose structure, and the water-saturated sand transferred into the liquefied state. The weight sank, the piezometers registered an increase of pressure in the water with a slow decrease back to the original value later. The level of sand in the vessel

196 *Sketches of several dilatancy manifestations*

also decreased due to a more dense grain repacking.

Similar phenomena occurring in natural conditions were described by Ryzhov & Vikharev (1959). A loose-filled embankment of fine alluvial sand, which was constructed for a rail and automobile road at the Kremenchug hydrostation building site (Ukraine) was subjected to an induced flowing during its flooding by means of a 1.5 m pole being pressed into its slope three times, close to the water level.

The first test failure spread for 14 m into the depth of the 4-meter-high embankment. The soil diffused 24 m outside the embankment boundaries, having covered 740 square meters. Flowing affected 330 m² of the embankment surface. The site received the sloping equal to 1:10-1:12, the original value being 1:1. All in all, up to 1000 m³ of soil was diffused.

The second test failure occurred after abrupt submerging of the pole into the embankment in the place where its surface part was 2.9 m high, and its underwater part was about 1 m high. The process of liquefaction over the area of 520 m² went on much the same as in the first case, and spread for 19.2 m into the depth of the embankment. The liquefacted soil sloping at 1:10-1:12 occupied the area 1100 m² and overflowed beside the embankment boundaries for about 28 m. Approximately 1500 m³ of soil were affected by flowing. An accidental liquefaction case was registered at the same locality when the embankment was by chance struck by the oar of a passing boat.

The phenomenon of sand liquefaction is treated in a detailed way in the works of Yaropolsky (1933) and Casagrande (1935). The authors independently from one another published the results of their experiments which revealed the fact that the volume of sand in the shear area changes. In the case of dense composition the sand grains loosen during the sliding, whereas if the sand is loose the shear causes failure of its unstable structure with a possible partial loss of contacts.

During the shear of water-saturated sands there appear filtration phenomena influencing the stability of bases and slopes. The stability of non-cohesive soils having a rigid skeleton is determined by friction of grains from the soil masses lying above, external load and capillary forces. In the process of loose sand shear, we shall use the soil model proposed by Gersevanov & Polshin (1948) which represents soil as a system consisting of balls of regular form (Fig. 12.9). Porosity between the balls in Fig. 12.9a will constitute 48%, coefficient $e = 0.91$; the respective values in Fig. 12.9b are 26% and 0.35. Height drastically changes when the system is skewed. The water permeability will change by an order of magnitude (Sobolevsky Yu. A., 1975).

Low-humid sand shear tests conducted in dilatometric apparatuses of direct shear and internal bulge (DTA) allow to reveal the dependence of resistance to shear of the density of composition, and at the same time to measure normal stresses caused by dilatancy in the shear zone (see Chapter 4). When shear area is localized in a limited volume inside the soil massif, resistance to shear is subject to the function

$$\tau_u = (\sigma_{n_o} \pm \sigma_d) \, \text{tg} \, \varphi' \tag{12.13}$$

where '−' corresponds to the case of unloading initial normal pressure σ_{n_o} when grain contact is temporarily lost during contraction.

When $\sigma_{n_o} = \sigma_d$, the soil in the contraction band is totally devoid of friction, even if the soil is not suspended in water. A slit appears in the soil massif. Even the pore air can become the suspending medium, but most often sand is suspended by porous water.

The layer of soil, compacted in the process of contraction, like a half-permeable membrane transfers pressure from the adjacent massif (still not violated) to the emerged water interlayer. In this interlayer grains partially or completely loose contacts, and, as a result, porous pressure rapidly increases (Fig. 12.10). The surplus head causes upcoming filtration water current, and this current destroys the unstable structure of loose water-saturated sand. The contracting layer also suffers failure. Due to the instability of the soil structure the contraction phenomena and the membrane effect develop in the adjacent areas. The process acquires an avalanche rapidity, and liquefaction of considerable soil masses takes place. Constructions built on such soils submerge, natural and artificial slopes

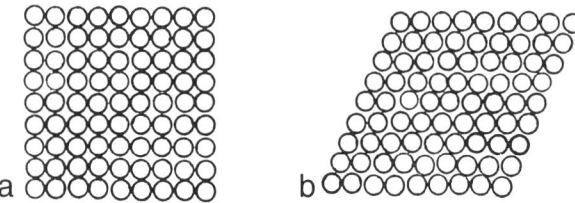

Figure 12.9. Mechanical ball models of ideal soil: a) unstable scheme with six contact points; b) stable scheme with eight contact points.

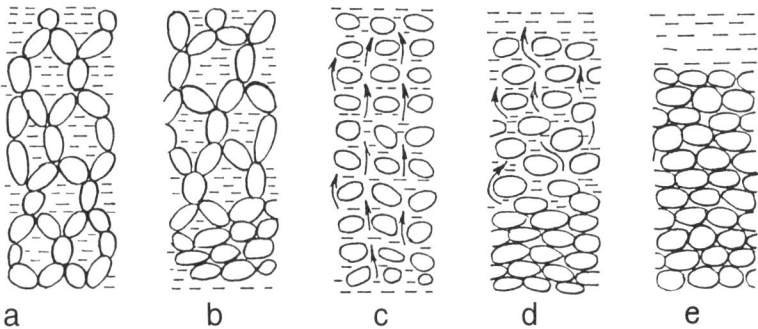

Figure 12.10. Grain repacking in loose water-saturated sand: a) original loose packing; b) repacking of grains of the lower contracting layer; c) suspension of grains by the upcoming filtration flow; d) process of particle repacking; e) compacted grain packing.

flow down. The liquefaction and flowing of water-saturated soils stops only when pressure in the pore water decreases and grain contacts are restored, i.e. when the aggregate friction at grain contact points ensures the preservation of a stable structure.

Full and partial sand liquefaction are distinguished (Ivanov, 1985). In the case of full liquefaction the loss of contacts spreads upwards from the contracting area, reaching the upper surface of the soil. Soil turns into suspension:

$$\gamma_{susp} = \gamma_s m + \gamma_w n \qquad (12.14)$$

where γ_s and γ_w are, respectively, specific weights of grains and water; n, m are, respectively, the shares of pores and skeleton in a unit of soil volume.

Pressure in the water at the level of the emerged slit under the conditional membrane increases to

$$P = \gamma_{sr} h = \gamma_{susp} h \qquad (12.15)$$

where $\gamma_{sr} = \gamma_{susp}$ are, respectively, specific weights of water-saturated soil and suspension; h is the depth of the considered level.

Taking into account the fact that with increased pressure the water from the slit will tend to go upwards, while sand grains will move downwards, there appears the head

$$H = \frac{\gamma_{sr}}{\gamma_w} h \qquad (12.16)$$

Maximum gradients of the head can reach $i = H/h$, i.e. become over 1. At the same time the sand mass is liquefied. But this state also occurs when $i = \gamma_{sb}/\gamma_w$, then

$$u_{max} = (\gamma_{sr} - \gamma_w)(h - z) = \gamma_{sb}(h - z) \qquad (12.17)$$

where u_{max} is the porous pressure corresponding to complete liquefaction of soil, γ_{sb} is the specific weight of soil suspended in water.

The upcoming flow of water results in total loss of contact in the layer lying above the contracting area (Fig. 12.10).

One should keep in mind that surplus pressure during dynamic effects may fail to reach the maximum value; then a partial lightening of contacts takes place, and the soil retains its bearing capacity to a certain extent. Ivanov (1985) suggested that such a condition should be assessed through the degree of soil liquefaction,

$$N = \frac{u}{u_{max}} = \frac{u}{\gamma_{sb}(h - z)} \qquad (12.18)$$

where u is the surplus porous pressure during partial liquefaction.

Numerically the values of N fall within the range from 0 to 1. When $N = 1$ it signifies a total liquefaction of soil.

In the process of partial liquefaction of water-saturated sands (which also leads

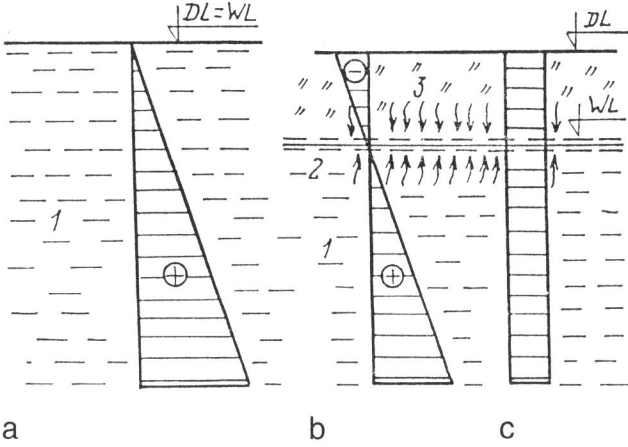

Figure 12.11. Emerging of capillary compressing forces in the process of repacking of loose finegrained water-saturated sand: a) distribution of stresses in pore water of loosely compacted sand; b) distribution of stresses in pore water of compacting sand; c – stresses in the sand skeleton due to capillary compression. 1. free water ('+' means that pressure in the water is higher than in the atmosphere); 2. capillary water ('-' means that pressure in the water is less than in the atmosphere); 3. dilating interlayer.

to grain repacking and compaction) dilatant phenomena, contrary to contraction, come into play (Fig. 12.11). Dilatancy of water-saturated sand is accompanied by loosening of the interlayer between the shearing soil areas with porosity exceeding the critical value. Rarefaction occurs. Water from the shearing parts of soil 'rushes' into the dilating interlayer. If the soil mass consists of sands with fine enough grains, a capillary water condition appears above the horizon 2-2, and the skeleton of the soil received capillary compression all through the soil mass (Vilun, 1976). This corroborates the conclusion made by Gersevanov & Polshin (1948) to the effect that '...sand with porosity below the critical value represents one of the most reliable bases'.

A comparison of conditions of contraction and dilatancy reveals that partial liquefaction of sands takes place primarily in soil masses whose structure is close (in terms of composition) to critical density, or when the massif consists of alternating loose and dense interlayers. Vibro-creeping of sands due to dynamic effects, apparently, can also be treated as dilatancy and contraction alternating in time and space.

Distortion notions about soil deformations, uniting dilatancy and contraction, point out the fact that drainage is effective for elimination of head in porous water, and that it is necessary to compact water-saturated sands till they reach the density exceeding the critical value in order to prevent or limit contraction and, consequently, liquefaction of sands.

12.3 DILATANT NATURE OF CONTACT FILTRATION AND NEGATIVE FRICTION ALONG THE PILE SHAFTS

The use of long piles has put on the research agenda the question of taking into account the negative friction. This phenomenon is manifested when part of the soil massif settles relative to the piles as a result of construction site planning (including increasing territory marks, loading the soil, change of soil weight following a decrease of subterranean water level or consolidation, and in some other cases) (Dalmatov et al., 1975).

Below we consider the cases of settlement due to contact filtration and negative friction during the adherence of soil onto along pile shafts. It is assumed that the uphanging and bearing layers are represented by friction (skeleton) soils, whose resistance to shear is determined by mechanical friction during mutual grain gear.

Let us consider the work conditions of piles represented in Figs 12.12 and 12.13 basing ourselves on the models set forth in Section 2.2, and the regularities of dilating soil behaviour.

The driving of pile causes the appearance in the soil massif of thrust stresses. During the loading of the pile, it is pressed further in and a contact layer is formed on the shaft surface. Grain packing in this contact layer in the process of shear settlement tends to get into the condition of critical density. In this way, along the contact surface of the pile there form the conditions for water drainage from the underlying non-consolidated soil. To justify the possibility of similar phenomena we can make use of the Coseni formula for establishing the coefficient of filtration of sandy soil:

$$k_f = 7.94 \frac{e^3}{1 - e^2} t d_{50} \qquad (12.19)$$

where e is the coefficient of soil porosity; t is the temperature coefficient; d_{50} is the mean grain diameter, mm.

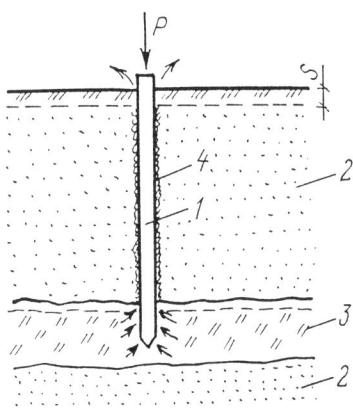

Figure 12.12. Pile hanging up in a weak soil layer: 1. pile; 2. bearing mass; 3. weak layer; 4. contact draining layer.

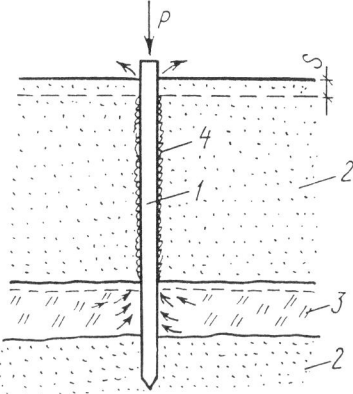

Figure 12.13. Pile passing a layer of weak non-consolidated soil: 1. pile; 2. bearing massif; 3. weak layer; 4. contact draining layer.

Let it be noted that when the porosity coefficient increase of 0.45 to 0.70 (say, for medium-sized sand), the value of filtration coefficient will increase by six times, while for friction clayey soil (e.g. moraine loam) such an increase can be tantamount to an order of magnitude and even more.

Therefore, dilatancy of the contact layer helps to mobilize resistance to shear, on the one hand, and forms around the pile a potentially draining layer, on the other hand. The possibility of slow consolidation of the weak underlying layer, arising as a result of this, is capable of gradually changing the work conditions of the pile. These conditions are different for the cases represented in Figs 12.12 and 12.13. Let us consider them.

Case 1. Pile hanging up in a layer of non-consolidated weak soil undergoes loading F (Fig. 12.12). Gradual displacement of water by the weight of the bearing mass with contact filtration along the pile surface leads to a slow change of the volume of the weak underlying soil. A settlement of the surface together with the pile takes place. The speed of such settlement can be calculated if we consider the pile field as a system of vertical drains, and the bearing mass as the draining additional weight.

Case 2. The pile passes through the layer of frictional and weak plastic water-saturated soil and its lower end reaches the bearing massif (Fig. 12.13). The bearing capacity of such a pile during the initial period is ensured by the resistance of soil along the contact surface of the shaft and under the pile.

As it was demonstrated above, dilatancy of the contact layer, along with the mobilization of resistance to shear, creates conditions or draining water from the non-consolidated weak layer. A slow settlement of the upper and consolidating soil layers takes place. The process can be further intensified when additional load is applied to the soil surface, but in this case the pile has no possibility to move simultaneously downwards, as in Case 1.

After compression of the weak layer and when a surface settlement of several millimetres is achieved, a reverse-direction resistance to shear is mobilized, and

202 Sketches of several dilatancy manifestations

the upper soil mass adheres to the pile. As a result of this, increasing forces of negative friction are added to the active useful load, which cause a gradual transfer of all the load to the pile's point. This can lead either to additional settlement, or to the breaking of the shaft. The latter is especially dangerous in the case of long compound piles.

The above-discussed peculiarities of pile-behaviour do not manifest themselves during post-drive pile-tests. In both cases the measured bearing capacity can initially be quite sufficient for receiving the working load, while the phenomena of contact filtration and negative friction reveal themselves after a time (sometimes rather a long time), when the erection of the given construction is over. As a result, the newly-appearing settlements are unexpected, and their reasons are very hard to diagnose and interpret at once.

12.4 REALIZATION OF THE FACTOR OF DILATANT STRENGTH DURING TUNNELLING

Beginning with 1968, after the first successful employment in Frankfurt on the Main (Germany), the so-called new-Austrian tunnel method (NATM) received wide acclaim in tunnel construction and mining engineering. The peculiarity of this new method in comparison with traditional solutions for lining is the use of thin shotcrete layers instead of massive concrete of cast-iron tubing or monolitly pressed casing.

Tunnel driving with NATM use is effected in the following way. In the process of soil mining reinforcing nets are fixed on the soil surface; then the nets are

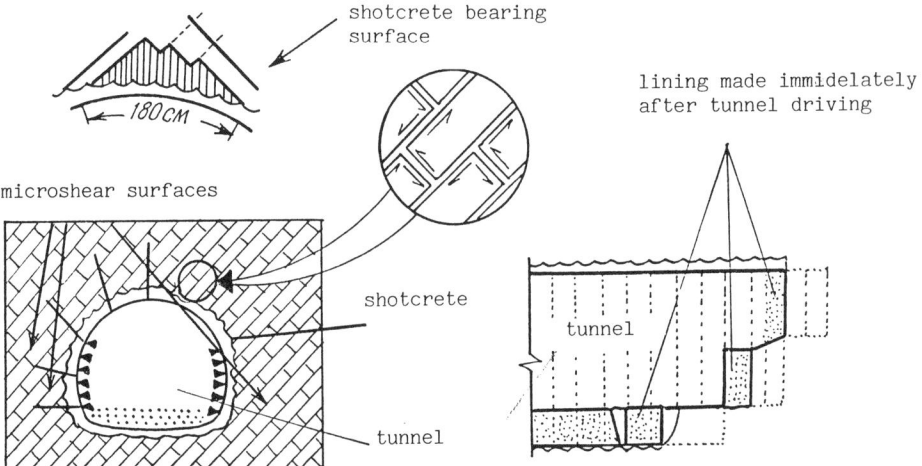

Figure 12.14. Tunnel lining with the use of NATM (Sauer, 1990): 1. tunnel; 2. shotcrete; 3. anchor bolts.

immediately covered with a shotcrete layer. This layer quickly acquires strength, which ensures a reliable initial lining. Additional lining of the vault by short anchors, bolts etc., (Fig. 12.14) is provided for non-reliable soils (cracked rocks and sands).

As the experience of NATM use clearly demonstrates for various soil conditions, such tunnel lining is rather effective and reliable compared with traditional methods (Sauer, 1990).

Let us analyze work conditions of soil in the case of NATM lining in comparison with the orthodox tubing method.

Traditionally, tunnel driving is done while the face soil is excavated and brought away, with consequent manufacturing of circular lining in the form of tubing. In this case there remains a gap behind the ring, which is later filled with cement mortar. The pressure of soil on the lining increases with pouring and failure of the massif's composition above the vault, and at maximum it reaches the value of natural pressure. As a rule, it is accompanied by soil surface deformation and the formation of settlement mould.

In the NATM case the lining is done immediately after the soil excavation, it closely adheres to the tunnel walls, giving no possibility for local failures to develop. As a result, the initial stress condition is violated minimally and the change of the ratio of principal stresses is not so considerable. The minimal movements occurring in the soil mainly have the development of unwedging stresses, which quickly dampens shear movements and minor local movements.

With the use of nails or short anchors for additional lining of the vault, minimal movements additionally mobilize the development of unwedging stresses along their length. As a result, a reinforced soil massif compressed by dilatant thrust is formed over the vault.

Thus, the efficiency of NATM in unstable soils is a direct consequence of limiting plastic deformations of soil under its own weight.

12.5 DILATANCY AND SEISMIC ACTIVITY

Zones of increased seismic activity coincide with the places where the earth's crust crystallic slabs are joined. Violation of equilibrium in these places (slab movements) are manifested on the surface by earthquakes. Several models for these phenomena have been worked out by geophysics. Without attempting to create a new model we shall try to add some features taking into account the phenomenon of dilatancy.

Let us consider a place where two slabs are joined, when the slabs are actually divided by a crack. Formation of cracks is inevitably connected with soil crushing and breaking off of huge lumps of rock. That is why the edges of the crack are jagged and its space is filled with debris.

This system can remain balanced when the active stresses in the condition of volume compression have very high values. For instance, at depths of 1000-3000 m just the pressure caused by the weight of the overlaying rock can reach the values of 25-75 MPa. Considerably larger stresses are concentrated at slab joint contacts and debris contact points. A minimal movement of the slabs causes an increase of stresses at these contact points due to dilatancy, constrained by the elastic reaction of the monolithic rock. An accumulation of unwedging stresses takes place; this latter phenomenon is possible, though, only before the overcoming of the rock strength.

At the moment when pressure, corresponding to the rock strength, is achieved, a rock lump is broken off into the crack, the slab's edges are crushed etc. Dilatant thrust stresses are released in the form of monolithic slabs elastic relief; this causes the earth's crust tremor and the system comes into a new state of unstable equilibrium.

During big earthquakes after the first powerful tremors one can observe for quite a long time weak tremors which are a consequence of local freeing of dilatant thrust stresses at places of their concentration at slab joint contacts. When these local concentrations are 'discharged', the system passes into a more stable state and the seismic activity gradually fades away.

It appears that periodicity of earthquakes can be assessed from the time necessary to accumulate dilatant thrust stresses, while the appearance of tremors can be considered as the consequence of achieving by these stresses of mainland rock strength. We can suppose that the stronger risk manifests less considerable seismic activity, but the rare earthquakes turn out very destructive. And on the contrary, weaker rock can be more active seismically at the expense of frequent breaking off at joint contacts, but the tremors are much weaker.

The increase of seismic activity can often be accounted for by human actions. For example, creating artificial lakes/water reservoirs in deep gorges (canyons) rather often leads to the growth of earthquake frequency in the neighbouring regions. These earthquakes are the result of local violations of equilibrium at slab joints. The emerging subterranean tremors testify to the fact that the system tends to get into the state of equilibrium which would correspond to the new ratio of the active stresses. This is obstructed by the oscillation of water level in the reservoir, which may stretch the process of fading of seismic activity away in time.

12.6 DILATANCY IN CRACKS OF STONE CONSTRUCTIONS

Installation of steel tension bars is widely practised for strengthening deformed constructions of building with brickwork or stone masonry. Prestressed tension bars placed in toothings are especially effective. After their installations building very often can be successfully exploited for a long time; in some cases superstructures can be erected.

The efficiency of prestressed tension bars, beside increasing the overall rigidity

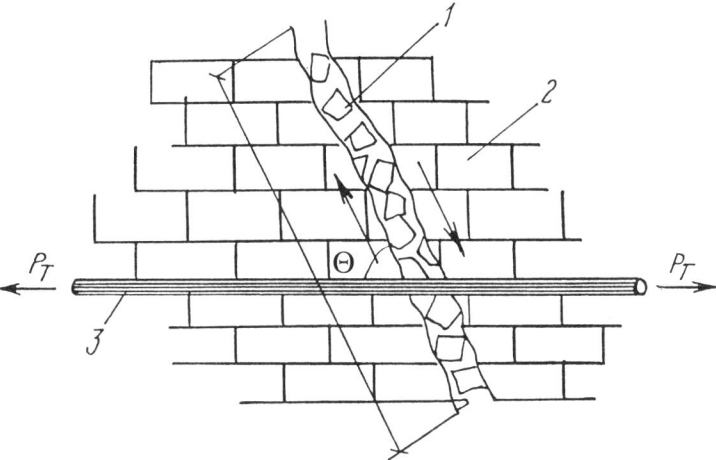

Figure 12.15. Scheme of a crack in a brick wall fixed by a tension bar: 1. crack with uneven surface filled with debris; 2. brickwork; 3. steel tension bar.

of the construction, is connected with the appearance of considerable friction forces in crack locations. Let us dwell on the mechanism of development of these forces.

Fig. 12.15 represents a crack in the brickwork of a wall of thickness B, crossed by a tension bar of cross section A, length l, prestressed force P_t. The crack has an uneven surface and is partially filled with debris. In the crack's plane at the length of opening L acts normal pressure

$$\sigma_n = \frac{P_t \cos \Theta}{BL} \tag{12.20}$$

If, as a result of additional settlement, one part of the wall shifts relative to another part, then dilatancy constrained by the rigidity of the tension bar will develop in the place of the crack.

Let us assume that dilatant movement in the crack constitutes Δl. The increment of force in the tension bar corresponding to this movement will be

$$\Delta P_t = \frac{\Delta l}{l} E_t A_t \tag{12.21}$$

where E_t is modulus of elasticity of the tension bar.

So, when elastic tie is installed, movement of one part of masonry relatively to another is immediately accompanied by an increase of normal pressure in the failure place. Consequently, resistance to shear also increases, and it can be calculated as

$$\tau = \frac{P_t + \Delta P_t}{BL} \operatorname{tg} \psi'_c \tag{12.22}$$

where ψ'_c is mobilized angle of contact friction.

Thus, if a crack is stopped by a tension bar, dilatant thrust in its plane turns into a strength factor. And vice versa, if dilatancy is not constrained in any way, it is directed to further opening of the crack, that is, becomes a failure factor.

An analysis of expression (12.21) reveals that, other things being equal, the efficiency of the dilatancy factor is in reverse proportion to the free length of the tension bar L_f. When the bar is grouted into the toothing this length is limited by only a short section near the crack. If no grouting is employed, the effect of prestress is rapidly decreased, especially for bars with length exceeding 4 to 6 m. Let us take an example to clarify this.

Dilatancy equal to 0.1 mm will cause in a rod (diameter 20 mm, length 6 m, modulus of elasticity of steel $E = 2.0 \cdot 10$ MPa) an increment of force equal to 1 kN, whereas if the bar is grouted (the free length near the crack is 0.2 m), the increment will constitute 30 kN, i.e. 30 times more. Thus, other things being equal, an increment of force due to dilatancy is in reverse proportion to the length of the elastic connection. Of course dilatant wedging manifests itself as a strength factor only if stresses developing in the crack do not exceed the strength of the masonry material.

It appears that the suggested dilatant approach can be useful for working out a theoretical calculation method of strengthening stone constructions. Such method would allow the optimization of rigidity of connections and their prestressing, refinement of constructive schemes, assessment of general stability etc.

12.7 REINFORCED CONCRETE AS A DILATING COMPOSITE MATERIAL

The mechanism of dilatancy described in Section 12.6 can also be useful while considering the work of concrete constructions. As an example we shall take characteristic cracks in the sections of beams adjacent to the support joints. These cracks appear as a result of crosswise force action. Slanting reinforcement oriented approximately perpendicularly to the possible crack direction is predesigned in the framework to receive this force. As a result, the minimum shear in this place is resisted by contact friction which increases due to the dilatancy of the filling material.

There are constructive schemes of reinforced concrete frameworks, where beams are fixed to columns without supporting tables, only at the expense of prestress of reinforcement on the concrete. The system remains stable due to the friction of the beam's end in the contact plane.

Reliability of such connection and the true value of the friction appearing therein can be assessed taking into account the dilatancy in the plane of potential shear. The friction forces developing with minimal shear can turn out to be so considerable that failure will take place outside the contact plane.

In prestress on concrete, injection of reinforcement channels is a major factor. The usefulness of such injection is analogous to placing a tension bar into a grouted toothing, i.e. it is determined by a decrease of free length of reinforcement at places adjacent to the planes of potential failure.

Taking into account the dilatancy of the filling material seems to be very useful in establishing the strength of concrete as composite material. For example, the possibility of prestressing concrete on filling materials of various strength can thus be evaluated.

Composite materials analogous to concrete consist of the filling material (aggregate) and reinforcement. The binding aggregate provides gear and creates an internal rigid structure. Keeping this in mind, materials of the concrete type can be regarded as systems characterized by internal friction and binding properties. Each contact of the aggregate in such systems is a potential area of friction and development of dilatancy (failure). This latter is suppressed by the strength of the binding and by reinforcement.

Fibroconcrete can serve as a symptomatic example of turning dilatancy from a failure factor into a factor of internal stability. The loading of this material causes local stresses of chaotically oriented small wires anchored in the binding. As a result, the dilatancy of the aggregate becomes elastically constrained in volume, and fibroconcrete behaves as an isotropic material which sustains both static and shock effects well (for example, in driven pile top).

General conclusions

1.1. The strength theory of Coulomb-Mohr in its traditional understanding is true only for the conditions of free dilatancy and the critical initial density of composition of grainy medium. This peculiarity is the main reason of failures when employing it for solving a number of tasks facing modern geotechnics. In order to adjust the conditions of strength to the conditions of constrained dilatancy, it is necessary to introduce two principal amendments:

a) Normal pressures in the process of failure of grainy medium are not constant, but depend on the reaction of the surrounding massif to the accompanying volume deformations;

b) The angle of internal (contact) friction depends on the degree of realization of the angle of grain gear, which is determined by the conditions of development of dilatancy.

Therefore, dilatancy can be taken as the physical property which connects strength and deformativity of the grainy medium.

1.2. Constrain of dilatancy leads to changes of initial normal pressures acting in the grainy medium as a result of appearance of thrust or during the repacking of grains of the shear surface or of the bulge area. Thrust dilatant stresses are in direct function to massif rigidity, soil density, grain size and strength of the soil-forming mineral. When shear occurs in a soil whose density is lower than initial critical density, grain gear partially disappears, contraction with relief of initial stresses takes place, resistance to shear decreases.

1.3. Constraint of dilatancy changes the character of dependence of resistance to shear of non-cohesive soil. Grain unwedging on the sliding surface leads to the appearance of a dilatant component of resistance to shear, whose value is determined by rigidity of the massif. The increase of this component approximates the mechanism of shear to the conditions of mutual sliding of solid elastic blocks. The angle of internal (contact) friction in the conditions of free dilatancy decreases in the direction of the value of the angle of intergranular friction and takes a certain intermediate value. Correspondingly, the Coulomb law takes the equations (4.1), (4.2) or (5.1).

1.4. Constraint of dilatancy is a reserve of additional soil strength during internal bulge. This is usually not taken into account by the existing notions. Unwedging of grains in the bulge area causes an increment of normal pressures by the value of dilatant pressures, and changes their ratio in limit state at the expense of a decrease of the angle of grain gear in the direction of the value of the angle of intergranular friction. The condition of strength during a triaxial compression takes the equation (6.3), whereas the maximum resistance to internal bulge can be represented in the equation (6.4) through initial lateral pressure in the massif.

1.5. Exclusion of constraint of dilatancy returns the conditions of strength during shear (4.2), (5.1) and during triaxial compression (6.3), to their usual equation, respectively, (1.3) and (1.4), which, consequently, is their specific case.

1.6. Traditional apparatus and test methods do not allow us to take into account the influence of dilatancy on strength, when there is resistance to volume deformations in soil, which is why there is a necessity to move on to the use of apparatus and methods enabling researchers to model real failure conditions. They should be designed on the basis of elastic-plastic models built on the sharp distinction between dilatant plastic deformations and elastic deformations which determine the former.

1.7. In the conditions of constrained dilatancy there is a close correlation between strength and deformative parameters of grainy medium. The described functions of the equations (4.7), (4.8) and (4.10), (4.11) open up possibilities for a mathematic description of strength as a quantity determined by physical properties and mechanical characteristics of grainy medium and by geometrical dimensions of the failure area: density of compositions, size and roughness of grains, humidity, modulus of elasticity and Poisson coefficient, size and form of the sliding surface or bulge area.

2.1. In the light of the revealed role of dilatancy in the strength of non-cohesive soil, technology of manufacturing deep foundations appears to be a factor which determines failure conditions corresponding to both the schemes of free and constrained dilatancy.

Traditional technologies of bore piles manufacture cause considerable soil violations, which determine uncertainty of conditions of mobilization of contact friction and resistance to internal bulge. This uncertainty together with a lack of understanding of peculiarities of free and constrained dilatancy makes impossible or hampers the creation of a theoretical calculation method. At the same time, technologies of injection compression ensure soil compaction with compensation of violations after its excavation and mobilizes resistance to contact shear and internal bulge in the conditions of constrained dilatancy with minimal settlements. This opens up possibilities for purposefully influencing the soil with the aim of reaching optimum conditions for realization of its potential strength.

2.2. Pile and footing manufacturing technologies with hollow-stem auger soil excavation, under the protection of bentonite suspension, and the diaphragm walling method allow us to exclude considerable soil violations and relief of stresses acting therein. With full compliance with the requirements of these technologies there is a possibility to ensure conditions of constrained dilatancy, i.e. optimum soil work after loading the foundation.

The factor of constrained dilatancy is the principal reserve of increasing the bearing capacity and efficiency of deep foundations. Realizing it opens the road for perfecting the existing and working out new technologies ensuring active and purposeful influence on the base soil.

2.3. Employing the refined Coulomb-Mohr strength conditions brings closer the theoretically calculated and experimental values of load-holding capacity of piles manufactured with the use of different technologies.

A major advantage of piles manufactured with minimal soil violation, or with compensation of such violations by injection, is that their work responds to a precise design scheme. Compacting by injection the soil layer under the lower end of the pile (which was loosened in the process of excavation) ensures mobilization of resistance to internal bulge with settlements approximately the shear settlement.

The Coulomb-Mohr strength conditions amended for the conditions of constrained dilatancy can serve as a basis for creation of a calculation method for establishing the bearing capacity of footings constructed by the diaphragm walling method, and for optimizing their forms and dimension.

2.4. Resistance to contact shear developing in the conditions of constrained dilatancy in injection-compacted non-cohesive soil is the principal source of the load-holding capacity of injection anchors. Mean values of resistance to shear depend on cross and lengthwise deformations of the anchor root when it is loaded, as a result of their influence on the dilatancy of contact layer.

2.5. The use of strength conditions of dilating non-cohesive soils opens up possibilities for the analysis of reinforced soil as a composite material, and also clarifies the treatment of such phenomena as liquefaction of sands, negative friction along pile shafts etc.

Appendix 1

Table A1.1. Values of coefficient of form of shear surface according to Tsitovich (1963).

Form of the loaded surface, rectangles. ($\eta = l/b$) m^2	Value of the coefficient ω	Form of the loaded surface, rectangles. ($\eta = l/b$) m^2	Value of the coefficient ω
1	0.95	8	2.12
1.5	1.15	9	2.19
2	1.30	10	2.25
3	1.53	20	2.64
4	1.70	30	2.88
5	1.83	40	3.07
6	1.96	50	3.22
7	2.04	100	3.69

Table A1.2. Experimental and calculated values of ultimate dilatant movements.

Sand	K MN/m^3	σ_{n_o} kPa	Values of δ_d^{exp}, δ_d^{cal} and δ_d^{exp}/d_{50} with I_D equal to								
			1.0			0.8			0.6		
			δ_d^{exp}	δ_d^{cal}	δ_d^{exp}/d_{50}	δ_d^{exp}	δ_d^{cal}	δ_d^{exp}/d_{50}	δ_d^{exp}	δ_d^{cal}	δ_d^{exp}/d_{50}
Coarse (d_{50} =1.50 mm)	44	0	0.45	1.14	0.76	–	0.82	0.55	0.19	0.52	0.35
		100	0.36			0.15			0.04		
		200	0.18			0.05			0.03		
	1208	0	0.13	0.24	0.16	–	0.17	0.11	0.09	0.13	0.09
		100	0.11			0.10			0.05		
		200	0.07			0.06			0.04		
	2845	0	0.07	0.16	0.11	–	0.12	0.08	0.07	0.08	0.05
		100	0.06			0.04			0.06		
		200	0.05			0.03			0.05		
Medium sand (d_{50} = 0.50 mm)	44	0	0.37	1.32	2.64	–	0.98	1.96	0.21	0.61	1.22
		100	0.20			0.16			0.11		
		200	0.20			0.08			0		
	1208	0	0.10	0.16	0.32	–	0.10	0.20	–	0.06	0.12
		100	0.08			0.06			0.01		
		200	0.06			0.04			0		
	2845	0	0.05	0.09	0.18	–	0.06	0.12	–	0.04	0.08
		100	0.04			0.03			0.01		
		200	0.02			0.01			–0.01		

Appendix 1

Table A1.3. Values of angles of intergranular friction φ_μ measured by different authors (according to Rowe, 1972).

Apparatus, condition of soil	Author	Grain material	Mean grain diameter, mm	Mean pressure kN/m²	Pore liquid or condition on the grain surface	φ_μ, grades	φ_{max}, (φ_o) grades
Direct shear Free particles	Rowe, 1964	Steel	2.4	350	Air	7	
		Glass balls	0.25	350	Water-saturated	9	
						17	
			Large-grained sedimentary soil	350	Water-saturated	30	
			Medium-grained sand	350		28	
			Coarse sand	350		23	
The same	Hiat, 1965	Zircon	0.2	200		23	
	El-Sohby, 1969	Feldspar		200		36	
	Tong, 1970	Quartz		200		26	
			0.1–0.6		Water-saturated	15	
			0.1		Humid and water-saturated	9	
			0.1		Dry nitrogen test		
			0.1		Dry nitrogen test		
		Glass balls			Acetone-cleansed	16	37
		Glass balls	Dry nitrogen test		Trichlorethylene, acetone, cleansing		
					Water-saturated	21	46
					Cleansed with soap, water and acetone	15	
						15	
Apparatus of direct shear Particles waxed together after initial sliding			3		Cleansed with soap, water and acetone	15	
			5			14	

Table A1.3. Continued.

Apparatus	Author	Grain material	Mean grain diameter, mm	Mean pressure kN/m^2	Pore liquid or condition on the grain surface	φ_μ, grades	φ_{max}, (φ_o) grades
Sliding of fixed particles along flat surface	Tschebotarioff, 1948	Quartz	5	20	Dry	6	
The same	Lenman, 1953	Flat crystals	High loads	870	Moistened and water-saturated	25	
					Dry	11	
		The same			Water-saturated	33	
			Low loads			19	
	Bishop, 1962 Horn, 1962	Pure quartz	Pebbles		Air-dried	29	
					Dry	22	
					Water-saturated	6	
	'_'	White quartz			Dry	23	
					Water-saturated	9	
	'_'	Pink quartz			Dry	27	
					Water-saturated	7	
	'_'	Feldspar (microwedge)			Dry	24	
					Water-saturated	6	
	'_'	Transparent quartz			Dry	37	
					Amylamin	31	
	Bromwell, 1966				Carbontetrachloride	11	
					Atmospheric conditions	28	36
					High vacuum	38	42
Ball-against-ball sliding	Skinner, 1969	Glass balls	1	6*	Dry	2	
			3	52		4	
	'_'	The same	1	6	Immersed into water	28	
				52		38	
	'_'		3	6	The same	40	
				52		40	
Ball-against-surface sliding	'_'	Glass balls	3	10	Dry	3	
				90		7	
	'_'	The same		10	Immersed into water	37	
				90		40	

* Contact load in grams.

Table A1.4. Resistance of control-mix sands to triaxial compression in the conditions of free dilatancy and values of φ_o.

Sand	Index of density I_D	Lateral pressure σ_{03} = const, kPa	Max. vertical pressure σ_{01}, kPa	Value of φ_o, grades
Coarse (d_{50} = 1.50 mm)	0.8	50	250	41.8
		100	501	41.9
		200	1010	42.0
				mean 41.9
	0.6	50	240	40.7
		–	–	–
		200	950	40.7
				mean 40.7
Medium (d_{50} = 0.50 mm)	0.8	50	190	35.7
		100	380	35.7
		200	748	35.3
				mean 35.6
	0.6	50	184	34.9
		100	367	34.8
		200	728	34.7
				mean 34.8
Fine (d_{50} = 0.25 mm)	1.0	50	184	34.9
		100	349	33.7
		200	732	34.8
				mean 34.5

Table A1.6. Calculated values of coefficients of elastic resistance K for mean values of modulus of elasticity E = 356 MPa (in the numerator) and 232 MPa (in the denominator) depending on the form and ratio of dimensions of shear surface.

Radius or breadth of shear surface r, b m	Values of K, MPa with E = 356 and 232 for sums					
	Axis symmetrical	Flat with side ratio l/b (coefficient ω)				
		1(0.93)	5(1.83)	10(2.23)	50(3.22)	100(3.69)
0.1	2738/1785	4118/2684	2138/1393	1739/1133	1214/792	1060/694
0.2	1369/892	2058/1342	1069/697	869/567	607/396	530/345
0.5	548/357	824/537	428/279	348/227	243/158	212/138
1.0	274/178	412/268	214/139	174/113	121/79	106/69
1.5	182/119	275/179	143/93	116/76	81/53	71/46
2.0	137/89	206/134	107/70	67/57	60/40	53/35
2.5	110/71	165/107	86/56	70/45	49/32	42/28
3.0	91/59	137/89	71/46	58/38	41/26	35/23
5.0	55/36	82/54	43/28	35/23	24/16	21/14
10.0	23/18	41/27	22/14	17/11	12/8	11/7

Appendix 1 215

Table A1.5. Mean values of control-mix sands' resistance to triaxial compression in the conditions of constrained dilatancy.

Sand	Index of density I_D	Coefficient of elastic resistance K, MN/m^3	Initial (hydrostatic) pressure $\sigma_{03} = \sigma_{01}$, kPa	Values of mobilized principal stresses, kPa		Maximum increment of volume (dilatancy) ΔV_d, cm^3	Value of φ', grades
				σ_1	σ_3		
Coarse (d_{50} = 1.50mm)	0.8	70	0	266	51	1.82	42.7
			50	518	100	1.79	42.6
			100	760	150	1.79	42.1
			200	–	–	–	–
							mean. 42.4
		1933	0	630	190	0.25	32.4
			50	791	240	0.25	32.3
			100	941	291	0.25	31.8
			200	1199	379	0.36	31.3
							mean. 32.0
		4552	0	771	270	0.15	28.8
			50	1011	319	0.15	31.3
			100	1241	440	0.19	28.4
			200	1342	471	0.15	28.9
							mean. 29.3
	0.6	70	0	173	36	1.28	40.9
			50	407	84	1.21	41.1
			100	641	131	1.11	41.3
			–	–	–	–	–
							mean. 41.1
		1933	0	430	130	0.17	32.4
			50	587	181	0.17	31.9
			100	745	218	0.15	33.2
			200	1039	315	0.15	32.6
							mean. 32.5
		4552	0	540	178	0.098	30.3
			50	671	217	0.092	30.7
			100	818	268	0.092	30.4
			200	1100	366	0.091	30.1
							mean. 30.4
Medium (d_{50} = 0.50 mm)	0.8	70	0	97	26	0.93	35.3
			50	280	74	0.86	35.5
			100	471	124	0.86	35.6
			200	850	161	1.13	–
							mean. 35.5
		1933	0	260	86	0.11	30.2
			50	401	135	0.11	29.8
			100	550	180	0.10	30.5
			200	841	284	0.11	29.7
							mean. 30.0
		4552	0	340	120	0.066	28.6
			50	470	166	0.064	34.5
			100	611	219	0.065	28.1
			200	876	312	0.062	28.3
							mean. 29.8

Table A1.5. Continued.

Sand	Index of density I_D	Coefficient of elastic resistance K, MN/m³	Initial (hydro-static) pressure $\sigma_{03} = \sigma_{01}$, kPa	Values of mobilized principal stresses, kPa σ_1	σ_3	Maximum increment of volume (dilatancy) ΔV_d, cm³	Value of φ', grades
	0.6	70	0	60	15	0.54	36.9
			50	242	68	0.64	34.1
			100	431	116	0.57	35.2
			200	807	213	0.46	35.4
							mean. 35.4
		1933	0	161	54	0.070	29.8
			50	320	103	0.069	30.8
			100	460	158	0.075	29.3
			200	750	250	0.065	30.0
							mean 30.0
		4552	0	210	76	0.042	27.9
			50	–	–	–	–
			100	490	174	0.041	28.4
			200	760	270	0.038	29.0
							mean. 28.4
Fine (d_{50} = 0.25 mm)	1.0	70	0	136	38	1.36	34.3
			50	315	87	1.32	34.5
			100	–	–	–	–
			200	850	231	1.11	35.0
							mean. 34.6
		1933	0	360	120	0.16	30.0
			50	510	178	0.16	28.9
			100	630	221	0.16	28.7
			200	925	320	0.16	29.1
							mean. 29.2
	0.6	70	0	56	15	0.54	35.2
			50	230	65	0.54	34.0
			100	400	111	0.39	34.4
			200	708	191	0.32	35.1
							mean. 34.7
		1993	0	154	52	0.07	29.7
			50	–	–	–	–
			100	403	138	0.05	29.3
			200	712	244	0.06	29.3
							mean. 29.4
		4552	0	–	–	–	–
			50	301	102	0.03	29.6
			100	452	166	0.04	27.6
			200	698	255	0.03	27.7
							mean. 28.3

Appendix 1 217

Table A1.7. Calculated values of τ_d and φ'_c for coarse ($d_{50} = 1.50$ mm) dense ($I_D = 1.0$) sand depending on the form and ratio of dimensions of shear surface with $E = 356$ MPa (numerator) and 232 MPa (denominator).

Radius or breath of shear surface r, b m	Values of τ_d (kPa) and φ'_c (grades) for sums									
	Axis symmetrical		Flat with ratio l/b (coefficient ω)							
			1(0.95)		5(1.83)		10(2.25)		50(3.22)		100(3.69)	
	τ_d	φ'_c	τ_d	φ'_c	τ_d	φ'_c	τ_d	φ'_c	τ_d	φ'_c	τ_d	φ'_c
0.1	259/225	30/32	309/261	29/30	248/211	31/33	231/197	32/35	205/174	34/37	196/170	35/37
0.2	206/178	34/36	245/208	31/34	196/167	35/37	183/156	36/38	162/138	38/40	156/135	38/40
0.5	152/131	39/40	180/153	36/39	145/123	39/41	135/146	40/41	120/102	41/42	115/100	41/42
1.0	120/104	41/42	143/122	39/41	115/98	41/42	106/91	42/42	95/81	43/43	91/79	43/43
1.5	105/91	42/43	125/106	41/42	100/85	42/43	94/80	43/43	83/71	43/43	80/69	43/43
2.0	95/83	43/43	114/97	42/43	81/78	42/43	84/72	43/44	75/64	43/44	72/63	43/43
2.5	89/77	43/43	105/90	42/43	85/72	43/43	79/67	43/44	70/60	43/44	67/58	43/44
3.0	83/72	43/43	99/84	42/43	80/68	43/44	74/63	43/44	66/56	44/44	63/55	44/44
5.0	71/61	43/44	84/71	43/44	67/57	44/44	64/43	44/44	56/47	44/44	53/47	44/44
10.0	56/49	44/44	67/57	44/44	54/45	44/44	49/42	44/44	44/38	44/44	42/36	44/44

Table A1.8. Calculated values of τ_d and φ'_c for coarse ($d_{50} = 1.50$ mm) medium-dense ($I_D = 0.6$) sand depending on the form and ratio of dimensions of shear surface with $E = 356$ MPa (numerator) and 232 MPa (denominator).

Radius or breadth of shear surface r, b m	Values of τ_d (kPa) and φ'_c (grades) for sums									
	axis symmetrical		Flat with ratio l/b (coefficient ω)							
			1(0.95)		5(1.83)		10(2.25)		50(3.22)		100(3.69)	
	τ_d	φ'_c	τ_d	φ'_c	τ_d	φ'_c	τ_d	φ'_c	τ_d	φ'_c	τ_d	φ'_c
0.1	129/112	30/32	154/130	29/30	123/105	31/33	115/98	32/34	102/87	34/36	98/85	34/36
0.2	102/89	33/35	122/103	31/33	97/83	35/36	91/78	35/37	81/69	37/38	77/67	37/38
0.5	75/65	37/38	90/76	36/37	72/61	38/39	67/73	38/39	60/51	39/40	57/49	39/40
1.0	59/52	39/39	71/61	38/39	57/49	39/40	53/45	39/40	47/40	40/40	45/39	40/40
1.5	52/45	40/40	62/53	39/40	50/42	40/40	47/40	40/40	41/35	40/40	40/34	40/41
2.0	47/41	40/40	57/48	39/40	45/39	40/40	42/36	40/41	36/32	41/41	36/31	41/41
2.5	44/38	40/40	52/45	40/40	42/36	40/41	39/33	41/41	35/30	41/41	33/29	41/41
3.0	41/36	40/41	49/42	40/40	40/34	41/41	37/31	41/41	33/28	41/41	31/27	41/41
5.0	35/30	41/41	42/35	41/41	33/28	41/41	31/26	41/41	28/23	41/41	26/23	41/41
10.0	28/24	41/41	33/28	41/41	27/22	41/41	24/21	41/41	22/19	41/41	21/18	41/41

Table A1.9. Calculated values of τ_d (kPa) and φ'_c for medium-grained ($d_{50} = 0.50$ mm) dense ($I_D = 1.0$) sand depending on the form and ratio of dimensions of shear surface with $E = 356$ MPa (numerator) and 232 MPa (denominator).

Radius or breadth of shear surface r, b m	Values of τ_d (kPa) and φ'_c (grades) for sums										
	Axis symmetrical		Flat with ratio l/b								
			1		5		10		50		100	
	τ_d	φ'_c	τ_d	φ'_c	τ_d	φ'_c	τ_d	φ'_c	τ_d	φ'_c	τ_d	φ'_c
0.1	145/126	28/29	173/146	28/28	139/118	29/30	130/110	29/31	115/98	30/32	110/95	31/32
0.2	116/100	30/31	137/117	29/30	108/94	31/32	103/88	31/33	91/77	33/34	87/76	33/35
0.5	85/73	33/35	101/86	31/33	81/69	34/35	76/82	34/36	67/57	35/36	64/56	36/36
1.0	67/58	35/36	80/68	34/35	64/55	36/36	59/51	36/37	53/45	37/37	51/44	37/37
1.5	59/51	36/37	70/59	35/36	56/48	36/37	53/45	37/37	47/40	37/37	45/39	37/37
2.0	53/47	36/37	64/54	36/37	51/44	37/37	47/38	37/37	42/34	37/37	40/33	37/38
2.5	50/43	37/37	59/50	36/37	48/40	37/37	44/38	37/37	39/34	37/37	38/33	37/38
3.0	47/40	37/37	56/47	36/37	45/38	37/37	42/35	37/38	37/31	38/38	35/31	38/38
5.0	40/34	37/38	47/40	37/37	38/32	38/38	35/30	38/38	31/26	38/38	30/26	38/38
10.0	31/27	38/38	38/32	38/38	30/25	38/38	27/24	38/38	25/21	38/38	24/20	38/38

Table A1.10. Calculated values of τ_d and φ'_c for medium-grained ($d_{50} = 0.50$ mm) medium-dense ($I_D = 0.60$) sand depending on the form and ratio of dimensions of shear surface with $E = 356$ MPa (numerator) and 232 MPa (denominator).

Radius or breadth of shear surface r, b m	Values of τ_d (kPa) and φ'_c (grades) for sums										
	Axis symmetrical		Flat with ratio l/b								
			1		5		10		50		100	
	τ_d	φ'_c	τ_d	φ'_c	τ_d	φ'_c	τ_d	φ'_c	τ_d	φ'_c	τ_d	φ'_c
0.1	55/48	29/30	66/55	28/29	52/45	29/30	49/42	30/31	43/37	31/32	42/36	31/32
0.2	43/38	30/31	52/44	29/30	41/35	31/32	39/33	31/32	34/29	32/33	33/29	32/33
0.5	32/28	32/32	38/32	32/31	31/26	32/33	29/31	33/34	26/22	33/34	24/21	34/34
1.0	25/22	32/33	30/26	33/33	24/21	34/34	23/19	34/34	20/17	34/34	19/17	34/34
1.5	22/19	33/34	26/23	33/34	21/18	34/34	20/17	34/34	17/15	34/34	17/14	34/34
2.0	20/17	34/34	24/20	34/34	19/17	34/34	18/15	34/34	16/14	34/34	15/13	34/34
2.5	19/16	34/34	22/19	34/34	18/15	34/34	17/14	34/34	15/13	34/34	14/12	35/35
3.0	17/15	34/34	21/18	34/34	17/14	34/34	16/13	34/35	14/12	35/35	13/11	35/35
5.0	15/13	34/34	18/15	34/34	14/12	34/34	13/11	35/35	12/10	35/35	11/10	35/35
10.0	12/10	35/35	14/12	35/35	11/9	35/35	10/9	35/35	9/8	35/35	9/8	35/35

Appendix 1   219

Table A1.11. Calculated values of τ_d and φ'_c for fine (d_{50} = 0.25 mm) dense (I_D = 1.0) sand depending on the form and ratio of dimensions of shear surface with E = 356 MPa (numerator) and 232 MPa (denominator).

Radius or breadth of shear surface r, b m	Values of τ_d (kPa) and φ'_c (grades) for sums									
	Axis symmetrical		Flat with ratio l/b							
			1		5		10		100	
	τ_d	φ'_c	τ_d	φ'_c	τ_d	φ'_c	τ_d	φ'_c	τ_d	φ'_c
0.1	115/100	28/29	137/116	28/28	110/94	28/29	103/88	29/30	91/77	29/31	87/76	30/31
0.2	92/79	29/30	109/92	29/29	87/74	30/31	81/69	30/31	72/61	31/32	69/60	32/32
0.5	68/58	31/32	80/68	30/32	64/55	32/33	60/65	32/33	53/45	33/34	51/44	33/34
1.0	53/46	33/34	64/54	32/33	51/44	33/34	47/40	34/34	42/36	34/34	40/35	34/34
1.5	47/40	34/34	56/47	33/33	44/38	34/34	42/36	34/34	37/32	34/34	36/31	34/35
2.0	42/37	34/34	51/43	33/34	40/35	34/34	37/32	34/34	33/28	34/35	32/28	34/35
2.5	40/34	34/34	47/40	34/34	38/32	34/34	35/30	34/35	31/27	34/35	30/26	35/35
3.0	37/32	34/34	44/37	34/34	36/30	34/35	33/28	34/35	29/25	35/35	28/24	35/35
5.0	32/27	34/35	37/32	34/35	30/25	35/35	28/24	35/35	25/21	35/35	24/21	35/35
10.0	25/22	35/35	30/25	35/35	24/20	35/35	22/19	35/35	20/17	35/35	19/16	35/35

Table A1.12. Calculated values of τ_d and φ'_c for fine (d_{50} = 0.25 mm) medium-dense (I_D = 0.60) sand depending on the form and ratio of dimensions of shear surface with E = 356 MPa (numerator) and 232 MPa (denominator).

Radius or breadth of shear surface r, b m	Values of τ_d (kPa) and φ'_c (grades) for sums									
	Axis symmetrical		Flat with ratio l/b							
			1		5		10		100	
	τ_d	φ'_c	τ_d	φ'_c	τ_d	φ'_c	τ_d	φ'_c	τ_d	φ'_c
0.1	51/44	28/29	61/51	28/28	48/41	29/29	45/39	29/30	40/34	30/31	39/33	30/31
0.2	40/35	29/30	48/41	29/29	38/33	30/31	36/31	30/31	32/27	31/32	30/26	32/32
0.5	30/26	31/32	35/30	31/32	28/24	32/33	29/26	32/33	24/20	33/33	22/19	33/33
1.0	23/20	33/33	28/24	33/33	22/19	33/33	21/18	33/34	18/16	33/34	18/15	34/34
1.5	20/18	33/34	24/21	33/33	20/17	33/34	18/16	34/34	16/14	34/34	16/13	34/34
2.0	16/18	33/34	19/20	33/33	18/15	34/34	17/14	34/34	13/15	34/34	12/14	34/34
2.5	17/15	34/34	20/18	33/34	17/14	34/34	15/13	34/34	14/12	34/34	13/11	34/34
3.0	16/14	34/34	19/17	34/34	16/13	34/34	14/12	34/34	13/11	34/34	12/10	34/34
5.0	14/11	34/34	17/14	34/34	13/11	34/34	12/10	34/34	11/9	34/34	10/9	34/34
10.0	11/9	34/34	13/11	34/34	11/9	34/34	9/8	34/34	9/7	34/34	8/7	34/34

220 Appendix 1

Table A1.13. Calculated values of passive/active pressures of coarse (d_{50} = 1.50 mm) dense (I_D = 1.0) sand for the case of axis symmetrical sum.

Radius of shear surface r, m	Values of σ^{max} (numerator) and σ^{min} (denominator), kPa, with modulus of elasticity E and initial pressure $\sigma_{03(01)}$													
	356 MPa							232 MPa						
	$\sigma_{03(01)}$, kPa							$\sigma_{03(01)}$, kPa						
	0	100	200	300	500	700		0	100	200	300	500	700	
0.1	896/−300	1196/−260	1496/−232	1796/−199	2396/−133	2996/−66		810/−248	1135/−217	1460/−186	1785/−155	2435/−93	3085/−31	
0.2	775/−218	1129/−190	1483/−163	1837/−134	2545/−78	3253/−21		698/−178	1083/−152	1468/−126	1853/−100	2623/−48	3393/4	
0.5	611/−146	1013/−123	1415/−99	1817/−77	2621/−31	3425/15		561/−123	1021/−101	1481/−79	1941/−57	2861/−13	3781/31	
1.0	526/−113	1007/−91	1487/−70	1969/−50	2931/−10	3893/29		468/−94	972/−74	1476/−54	1980/−34	2988/3	3996/46	
1.5	473/−95	977/−75	1480/−55	1985/−34	2993/76	4001/45		419/−78	948/−59	1470/−40	2006/−21	3064/17	4122/55	
2.0	437/−82	966/−63	1495/−45	2024/−26	3082/12	4140/50		382/−71	911/−52	1440/−33	1969/−14	3027/24	4085/62	
2.5	409/−77	938/−58	1467/−40	1996/−21	3054/17	4112/55		354/−66	883/−47	1412/−28	1941/−9	2999/29	4057/67	
3.0	382/−71	911/−53	1440/−34	1969/−16	3027/18	4085/60		331/−62	860/−43	1389/−24	1918/−5	2976/33	4036/71	
Under free dilatancy	0/0	558/42	1116/84	1674/126	2790/210	3906/294		0/0	558/42	1116/84	1674/126	2790/210	3906/294	

Table A1.14. Calculated values of σ^{max} (numerator) and σ^{min} (denominator), kPa, with modulus of elasticity E and initial pressure $\sigma_{03(01)}$ of coarse (d_{50} = 1.50 mm) medium-dense (I_D = 1.0) sand for the case of axis symmetrical sum.

Radius of shear surface r, m	356 MPa							232 MPa						
	$\sigma_{03(01)}$, kPa							$\sigma_{03(01)}$, kPa						
	0	100	200	300	500	700		0	100	200	300	500	700	
0.1	446/−150	746/−116	1046/−82	1346/−48	1946/20	2546/88		403/−123	728/−92	1053/−61	1378/−30	2028/32	2678/94	
0.2	375/−110	714/−81	1053/−32	1392/−23	2385/35	3065/93		342/−93	711/−66	1080/−37	1448/−12	2187/42	2925/96	
0.5	302/−75	704/−50	1106/−25	1508/0	2312/50	3116/100		267/−64	687/−40	1107/−16	1527/8	2367/56	3207/104	
1.0	242/−57	685/−34	1128/−11	1568/12	2448/58	3328/104		218/−50	658/−27	1098/−4	1538/19	2418/65	3298/111	
1.5	223/−49	683/−27	1143/−5	1603/17	2523/61	3443/105		205/−42	659/−20	1113/2	1573/24	2493/68	3413/112	
2.0	188/−41	668/−19	1108/3	1368/25	2488/69	3408/113		163/−36	623/−14	1083/8	1543/30	2463/74	3383/118	
2.5	93/−39	594/−17	1095/5	1555/27	2475/71	3395/115		154/−33	637/−12	1120/12	1601/30	2563/72	3525/114	
3.0	0/0	481/46	962/91	1443/137	2405/228	3367/319		0/0	481/46	962/91	1443/137	2405/228	3367/319	
Under free dilatancy														

Table A1.15. Calculated values of passive/active pressures of medium ($d_{50} = 0.5$ mm) dense ($I_D = 1.0$) sand for the case of axis symmetrical sum.

Values of σ^{max} (numerator) and σ^{min} (denominator), kPa, with modulus of elasticity E and initial pressure $\sigma_{03(01)}$

Radius of shear surface r, m	356 MPa $\sigma_{03(01)}$, kPa							232 MPa $\sigma_{03(01)}$, kPa						
	0	100	200	300	500	700		0	100	200	300	500	700
0.1	481/−174	758/−138	1035/−102	1312/−66	1866/6	2420/78		428/−149	716/−114	1004/−79	1292/−44	1968/26	2444/96
0.2	401/−135	701/−100	1001/−66	1301/−32	1901/36	2501/104		354/−114	666/−82	978/−50	1290/−18	1914/46	2538/110
0.5	313/−92	652/−63	991/−34	1330/−5	2008/53	2686/111		285/−76	654/−49	1023/−22	1392/5	2130/59	2868/113
1.0	257/−70	626/−43	995/−16	1364/11	2102/65	2840/119		227/−58	621/−32	997/−6	1382/20	2153/72	2922/124
1.5	236/−59	616/−33	1001/−7	1386/19	2156/71	2926/123		205/−51	607/−26	1009/−1	1411/24	2215/74	3019/124
2.0	208/−53	593/−27	978/−1	1363/25	2133/25	2903/129		189/−47	591/−22	993/3	1395/28	2199/78	3003/128
2.5	201/−50	603/−25	1005/0	1407/25	2211/75	3015/126		173/−43	557/18	977/7	1379/32	2183/82	2987/132
3.0	189/−47	554/−22	993/3	1395/28	2199/78	3003/128		161/−40	563/−15	965/10	1367/35	2171/85	2975/135
Under free dilatancy	0/0	420/49	840/98	1260/147	2100/245	2940/343		0/0	420/49	840/98	1260/147	2100/245	2940/343

Table A1.16. Calculated values of passive/active pressures of medium ($d_{50} = 0.5$ mm) medium-dense ($I_D = 0.6$) sand for the case of axis symmetrical sum.

Values of σ^{max} (numerator) and σ^{min} (denominator), kPa, with modulus of elasticity E and initial pressure $\sigma_{03(01)}$

Radius of shear surface r, m	356 MPa $\sigma_{03(01)}$, kPa							232 MPa $\sigma_{03(01)}$, kPa						
	0	100	200	300	500	700		0	100	200	300	500	700
0.1	187/−65	475/−30	763/5	1051/40	1627/140	2203/180		166/−56	466/−22	766/12	1066/46	1666/114	2266/182
0.2	149/−50	449/−16	749/18	1049/52	1649/125	2250/188		135/−43	447/−11	759/21	1071/53	1695/117	2319/181
0.5	115/−35	440/−4	765/33	1090/67	1740/120	2390/182		104/−31	444/−2	787/27	1120/56	1798/114	2476/172
1.0	92/−27	437/−2	770/31	1109/60	1787/118	2465/176		83/−23	437/5	791/33	1145/61	1853/117	2561/173
1.5	83/−23	431/5	791/33	1145/61	1853/117	2561/173		71/−20	425/8	779/36	1133/64	1841/120	2549/176
2.0	75/−21	429/7	783/35	1197/63	1845/119	2533/175		64/−18	418/10	772/38	1126/66	1834/122	2542/178
2.5	71/−20	425/8	779/36	1133/64	1841/120	2549/176		61/−17	415/11	769/39	1123/67	1831/123	2539/179
3.0	63/−18	417/10	772/38	1126/66	1834/122	2542/178		56/−16	410/12	764/40	1118/68	1826/124	2534/180
Under free dilatancy	0/0	374/52	748/104	1122/156	1870/260	2618/364		0/0	374/52	748/104	1122/156	1870/260	2618/364

Table A1.17. Calculated values of passive/active pressures of fine ($d_{50} = 0.25$ mm) dense ($I_D = 1.0$) sand for the case of axis symmetrical sum.

Values of σ^{max} (numerator) and σ^{min} (denominator), kPa, with modulus of elasticity E and initial pressure $\sigma_{03(01)}$

Radius of shear surface r, m	356 MPa $\sigma_{03(01)}$, kPa						232 MPa $\sigma_{03(01)}$, kPa						
	0	100	200	300	500	700	0	100	200	300	500	700
0.1	382/−138	658/−102	934/−66	1210/−30	1762/42	2314/114	340/−118	628/−83	916/−48	1204/−13	1780/57	2356/127
0.2	313/−109	601/−74	889/−39	1177/−4	1753/66	2329/136	273/−91	573/−58	873/−25	1173/8	1773/74	2373/140
0.5	241/−78	553/−46	865/−14	1177/18	1801/82	2425/146	209/−64	534/−33	859/−2	1184/29	1834/91	2484/153
1.0	195/−57	534/−28	873/−1	1212/30	1890/88	2568/146	173/−49	527/−21	881/7	1235/35	1943/91	2651/147
1.5	177/−50	531/−22	885/6	1239/34	1947/90	2655/146	150/−42	504/−14	858/0	1212/42	1920/98	2628/154
2.0	158/−45	512/−17	866/11	1220/39	1928/95	2636/151	139/−39	493/−11	847/3	1201/45	1909/101	2617/157
2.5	150/−42	504/−14	858/14	1212/42	1920/98	2628/154	128/−36	482/−8	836/6	1190/48	1989/104	2606/160
3.0	139/−39	493/−11	847/17	1201/45	1909/101	2617/157	120/−34	474/−6	828/8	1182/50	1890/106	2592/162
Under free dilatancy	0/0	369/52	738/104	1107/156	1845/260	2583/364	0/0	369/52	738/104	1107/156	1845/260	2583/364

Table A1.18. Calculated values of passive/active pressures of fine ($d_{50} = 0.25$ mm) medium-dense ($I_D = 0.6$) sand for the case of axis symmetrical sum.

Values of σ^{max} (numerator) and σ^{min} (denominator), kPa, with modulus of elasticity E and initial pressure $\sigma_{03(01)}$

Radius of shear surface r, m	356 MPa $\sigma_{03(01)}$, kPa						232 MPa $\sigma_{03(01)}$, kPa						
	0	100	200	300	500	700	0	100	200	300	500	700
0.1	169/−61	445/−25	721/11	997/47	1549/119	2101/191	150/−5	538/−22	726/−13	1014/48	1590/118	2166/188
0.2	136/−47	424/−12	712/17	1000/49	1576/128	2156/198	121/−41	421/−7	721/27	1021/61	1621/129	2221/197
0.5	106/−34	418/−2	730/30	1042/62	1666/126	2290/190	94/−29	419/−2	744/33	1069/64	1719/126	2369/188
1.0	85/−24	424/5	763/34	1102/63	1780/121	2458/179	74/−22	413/7	752/36	1091/65	1769/123	2447/181
1.5	74/−21	413/8	752/37	1099/66	1769/124	2447/182	68/−19	420/9	776/37	1130/65	1838/121	2546/175
2.0	66/−19	405/10	744/39	1091/68	1761/126	2439/184	61/−17	415/11	769/39	1123/67	1831/123	2539/177
2.5	63/−18	417/10	771/38	1125/66	1853/122	2541/184	56/−16	410/12	764/40	1118/68	1826/124	2534/178
3.0	59/−17	413/11	767/39	1121/67	1849/123	2547/135	53/−15	407/13	761/41	1115/69	1823/125	2531/178
Under free dilatancy	0/0	358/53	716/106	1074/159	1790/265	2506/371	0/0	358/53	716/106	1074/159	1790/265	2506/371

Appendix 1   223

Table A1.19. Calculated values of passive/active pressures of coarse ($d_{50} = 1.50$ mm) dense ($I_D = 1.0$) sand for the case of flat sum.

Ratio of shear surface sides l/b	Breadth of the surface b, m	Values of σ^{max} (numerator) and σ^{min} (denominator), kPa, with E and $\sigma_{03(01)}$									
		356 MPa $\sigma_{03(01)}$, kPa						232 MPa $\sigma_{03(01)}$, kPa				
		100	200	300	500	100	200	300	500
1	0.5	1091/−157	1476/−131	1862/−105	2632/−53	1081/−123	1520/−100	1960/−78	2039/−32
	1.0	1139/−113	1478/−91	1918/−68	2797/−22	1017/−90	1498/−69	1979/−49	2942/−7
	2.0	1016/−81	1521/−62	2025/−42	3045/−2	975/−65	1504/−46	2032/−27	3090/−10
	5.0	950/−53	1505/−35	2060/−17	3170/18	$855/−43	1384/−24	1913/−5	2971/33
	10.0	870/−38	1425/−21	1980/−3	3090/33	823/−30	1378/−30	1933/5	3043/42
5	0.5	1047/−115	1487/−92	1926/−70	2805/−24	1021/−91	1502/−70	1984/−50	2947/−8
	1.0	986/−84	1467/−63	1949/−42	2912/9	944/−67	1449/−47	1953/−28	2962/12
	2.0	901/−51	1430/−32	1959/−14	3017/24	887/−49	1416/−30	1945/−11	3003/26
	5.0	870/−39	1425/−21	1980/−3	3090/33	823/−30	1378/−12	1973/5	3043/42
	10.0	809/−28	1364/−10	1919/8	3029/44	767/−20	1322/−2	1977/16	2987/52
10	0.5	1038/−104	1498/−92	1958/−60	2878/−17	1122/−112	1603/−91	2085/−71	3048/−29
	1.0	980/−74	1485/−55	1989/−35	2998/5	913/−61	1417/−41	1922/−21	2931/18
	2.0	915/−54	1444/−35	1973/−16	3331/21	860/−44	1389/−25	1918/−26	2976/32
	5.0	856/−36	1411/−18	1966/−3	3016/36	757/−18	1312/0	1867/17	2977/53
	10.0	785/−23	1340/−5	1895/12	3005/48	753/−17	1307/0	1862/18	2972/54
50	0.5	1008/−88	1489/−68	1971/47	2934/−5	962/−71	1468/−51	1971/−31	2980/8
	1.0	965/−64	1494/−44	2023/−26	3081/12	901/−51	1430/−32	1959/−13	3017/24
	2.0	873/−46	1402/−27	1931/−8	2989/29	856/−36	1411/−18	1966/0	3076/35
	5.0	818/−29	1373/−11	1928/6	3038/42	776/−22	1331/−4	1886/14	2996/50
	10.0	762/−19	1317/−1	1872/17	2982/53	734/−14	1289/4	1844/22	2954/58
100	0.5	986/−84	1467/−63	1948/−42	2912/−10	953/−69	1458/−49	1962/−29	2971/10
	1.0	947/−60	1476/−41	2005/−22	1063/15	892/−50	1421/−31	1950/−12	3008/25
	2.0	860/44	1389/25	1918/−6	2975/32	818/−36	1348/−17	1876/2	2934/39
	5.0	804/−27	1559/−9	1914/9	3024/45	776/−22	1331/−4	1886/14	2996/50
	10.0	753/−17	1307/0	1862/18	2972/54	724/−12	1279/5	1834/23	2944/59
Under free dilatancy		558/42	111/84	1674/84	2790/210	558/42	1116/84	1674/126	2790/210

Table A1.20. Calculated values of passive/active pressures of coarse ($d_{50} = 1.5$ mm) medium-dense ($I_D = 0.60$) sand for the case of flat sum.

Ratio of shear surface sides l/b	Breadth of the surface b, m	Values of σ^{max} (numerator) and σ^{min} (denominator), kPa, with E and $\sigma_{03(01)}$							
		356 MPa $\sigma_{03(01)}$, kPa				232 MPa $\sigma_{03(01)}$, kPa			
		100	200	300	500	100	200	300	500
1	0.5	738/−66	1123/−40	1508/−40	2279/38	707/−51	1109/−26	1511/−1	2316/48
	1.0	711/−45	1132/−22	1552/2	2393/50	695/−35	1134/−13	1574/10	2453/55
	2.0	678/−312	118/−9	1557/14	2436/59	666/−23	1125/−1	1585/20	2505/64
	5.0	666/−17	1147/3	1629/24	2592/65	635/−11	1116/9	1598/30	2561/72
	10.0	626/−9	1107/11	1589/32	2552/74	604/−5	1086/16	1567/37	2530/78
5	0.5	715/−46	1135/−22	1556/1	2397/47	595/−35	1135/−13	1574/10	2453/55
	1.0	678/−31	1118/−9	1557/14	2432/59	670/−24	1130/−2	1590/19	2509/63
	2.0	653/−20	1112/1	1573/23	2492/67	627/−14	1087/7	1546/29	2467/72
	5.0	626/−9	1107/11	1589/32	2552/74	604/−5	1085/16	1567/37	2530/78
	10.0	600/−4	1081/17	1563/38	2536/79	578/0	1059/21	1541/42	2504/84
10	0.5	695/−41	1115/−18	1536/6	2376/53	745/−47	1185/−24	1615/−1	2504/44
	1.0	687/−28	1147/−5	1607/16	2527/59	653/−20	1212/1	1572/23	2492/67
	2.0	649/−17	1099/4	1560/26	2479/69	6391/−12	1121/9	1602/29	2565/71
	5.0	617/−7	1099/13	1580/−34	2543/75	595/−3	1077/18	1558/39	2521/80
	10.0	587/−1	1068/19	1549/40	2512/82	573/1	1055/22	1536/43	2499/85
50	0.5	691/−34	1130/−12	1570/11	2449/56	678/−26	1138/−4	1598/17	2518/61
	1.0	661/−22	1121/0	1581/21	2501/65	631/−15	1091/6	1551/28	2471/71
	2.0	644/−13	1125/8	1607/28	2570/70	622/−8	1103/12	1585/33	2548/75
	10.0	578/0	1059/21	1541/43	2504/84	565/3	1046/24	1528/45	2490/86
100	0.5	678/−31	1118/−9	1557/14	2436/59	670/−24	1129/−2	1590/19	2509/63
	1.0	652/−20	1112/1	1573/23	2492/67	627/−14	1087/7	1547/29	2467/72
	2.0	639/−12	1121/9	1602/29	2565/71	617/−7	1099/13	1580/34	2543/75
	5.0	595/−3	1077/18	1558/38	2521/80	582/0	1064/20	1545/41	2508/82
	10.0	573/1	1055/22	1536/43	2499/85	650/4	1041/25	1523/46	2486/87
Under free dilatancy		481/46	962/91	1443/137	2405/228	481/46	962/911	1443/137	2405/228

Appendix 2

Table A2.1. Calculated mean resistances to contact shear $\bar{\tau}_s$ along the skin surface of bore piles of Types 2, 3, and additional pressures at the level of the lower end for control-mix sands.

Soil	Pile diameter d, m	Density of soil	Values of $\bar{\tau}_s$* and σ_{ad}, kPa, with pile length L, m											
			6		8		10		12		14		16	
			$\bar{\tau}_s$	σ_{ad}	$\bar{\tau}_s$	σ_{ad}	$\bar{\tau}_s$	σ_{ad}	$\bar{\tau}_s$	σ_{ad}	$\bar{\tau}_s$	σ_{ad}	$\bar{\tau}_s$	σ_{ad}
Coarse sand	0.5	Dense	183	230	194	244	204	257	214	270	224	282	235	296
		Medium	115	145	129	162	142	179	156	196	170	214	183	230
	1.0	Dense	167	210	182	230	197	248	212	267	228	289	245	309
		Medium	102	129	117	148	132	166	147	185	161	203	175	221
	1.5	Dense	153	194	170	214	186	234	202	255	218	275	235	296
		Medium	97	123	112	142	127	160	142	179	157	199	172	217
Medium-grained sand	0.5	Dense	120	151	132	166	144	181	156	197	168	212	179	225
		Medium	66	83	77	97	88	111	99	125	110	139	122	140
	1.0	Dense	105	132	117	148	129	163	143	180	155	196	168	211
		Medium	60	76	72	90	84	106	96	121	107	135	119	149
	1.5	Dense	98	123	111	140	124	156	137	173	150	189	164	206
		Medium	58	74	71	89	84	106	97	122	110	139	123	139
Fine sand	0.5	Dense	100	126	111	140	122	154	133	167	144	181	154	195
		Medium	62	77	73	92	84	106	95	120	106	133	116	147
	1.0	Dense	88	111	100	126	112	141	124	156	136	171	147	185
		Medium	58	73	70	88	82	103	94	118	106	134	117	147
	1.5	Dense	83	105	95	120	107	135	119	150	131	165	143	181
		Medium	55	69	67	84	78	98	89	112	102	129	114	143

* The calculation was made with $E = 356$ MPa.

Table A2.2. Calculated resistances at the lower end of bore piles of Types 2, 3, R_f for control-mix sands.

Soil	Pile diameter d, m	Density of soil	Values of R_f*, kPa with pile length L, m					
			6	8	10	12	14	16
Coarse sand	0.5	Dense	9450	10186	10908	11629	12335	13071
		Medium	5820	6537	7254	7970	8700	9404
	1.0	Dense	9344	10248	10084	11470	12277	13132
		Medium	5480	6295	7094	7908	8707	9506
	1.5	Dense	8878	9823	10768	11730	12657	13638
		Medium	5347	6196	6752	7586	8743	9577
	Mean	Dense	9225	10020	10790	11610	12425	13280
		Medium	5550	6349	7030	7820	8720	9495
Medium-grained sand	0.5	Dense	5445	6024	6603	7190	7772	8330
		Medium	3000	3544	4088	4632	5176	5578
	1.0	Dense	5034	5679	6311	6969	7613	8246
		Medium	2726	3394	3985	4564	5132	5700
	1.5	Dense	4844	5531	6205	6892	7566	8252
		Medium	2716	3391	3916	4510	5117	5529
	Mean	Dense	5110	5750	6730	7020	7650	8255
		Medium	2815	3410	3920	4570	5140	5600
Fine sand	0.5	Dense	4378	4902	5426	5940	6464	6988
		Medium	2788	3322	3847	4371	4884	5408
	1.0	Dense	4048	4627	5207	5786	6366	6934
		Medium	2733	3313	3892	4471	5062	5618
	1.5	Dense	3944	4550	5156	5761	6367	6967
		Medium	2600	3179	3747	4315	4917	5385
	Mean	Dense	4120	4690	5263	5830	6400	6965
		Medium	2700	3270	3830	4385	4950	5505

* The calculation was made with $E = 356$ MPa.

Table A2.3. Calculated mean resistances to contact shear $\bar{\tau}_s$ along the skin surface of bore piles with injected shaft (Type 4), and additional pressures at the level of the lower end σ_{ad} for control-mix sands.

Soil	Pile diameter d, m	Index I_D of soil density	Values of $\bar{\tau}_s$ and σ_{ad}*, kPa with pile length L, m											
			4	6		8		10		12		14		16	
			$\bar{\tau}_s$	$\bar{\tau}_s$	σ_{ad}	$\bar{\tau}_s$	σ_{ad}	$\bar{\tau}_s$	σ_{ad}	$\bar{\tau}_s$	σ_{ad}	$\bar{\tau}_s$	σ_{ad}	$\bar{\tau}_s$	σ_{ad}
Coarse sand	0.5	1.0	15	254	320	268	338	283	357	297	374	311	392	326	412
		0.8	14	218	275	228	287	239	301	250	315	261	329	272	343
	1.0	1.0	15	242	305	248	312	255	321	261	329	267	336	273	344
		0.8	14	179	226	193	243	207	261	221	278	235	296	249	314
	1.5	1.0	15	218	275	234	295	251	316	268	338	284	358	299	377
		0.8	14	203	255	218	274	233	293	247	311	263	331	277	349
Medium-grained sand	0.5	1.0	12	211	266	229	289	247	311	265	334	283	357	301	379
		0.8	11	148	186	160	202	172	217	185	233	197	248	209	263
	1.0	1.0	12	172	217	177	223	182	229	187	236	192	242	198	249
		0.8	11	140	176	152	192	166	209	176	221	189	231	200	252
	1.5	1.0	12	150	190	164	207	177	223	192	242	205	258	216	272
		0.8	11	134	169	148	186	162	204	174	219	186	234	199	251
Fine sand	0.5	1.0	11	144	181	154	194	164	207	175	221	187	236	198	250
		0.8	11	137	172	148	186	158	199	168	211	178	224	189	238
	1.0	1.0	11	135	170	146	185	158	199	169	213	182	229	193	243
		0.8	11	124	156	128	161	142	179	157	198	170	214	183	231
	1.5	1.0	11	131	165	143	180	156	197	168	212	180	227	192	242
		0.8	11	114	144	126	158	138	174	150	189	162	204	174	219

* The calculation was made with $E = 356$ MPa.

228 *Appendix 2*

Table A2.4. Calculated resistances along the injected lower end of bore piles of Types 4, 5, R_f for control-mix sands.

Soil	Pile diameter d_2, m	Index of soil density I_D	Values of R_f*, kPa, with pile length L, m					
			6	8	10	12	14	16
Course sand	0.5	1.0	11417	12216	13030	13815	14614	15443
		0.8	10017	10693	11397	12101	12805	13509
	1.0	1.0	10812	11500	12221	12925	13619	14331
		0.8	8958	9775	10610	11428	12262	13096
	1.5	1.0	10245	11190	12152	13132	14077	15005
		0.8	8891	9754	10617	11463	12340	13188
	Mean	1.0	10825	11635	12470	13290	14100	14920
		0.8	8290	10075	10875	11665	12470	13265
Medium sand	0.5	1.0	6751	7421	8080	8750	9421	10080
		0.8	5156	5740	6310	6892	7464	9005
	1.0	1.0	5889	6390	6890	7403	7904	8416
		0.8	4919	5539	6171	6743	7291	7971
	1.5	1.0	5712	6389	7073	7749	8419	9107
		0.8	4867	5536	6219	6864	7509	8210
	Mean	1.0	6120	6735	7350	7970	8580	9200
		0.8	4980	5605	6235	6835	7420	8395
Fine sand	0.5	1.0	4977	5495	6012	6541	7079	7606
		0.8	4740	5268	5786	6293	6811	7339
Fine sand	1.0	1.0	4720	5299	5867	6435	7026	7594
		0.8	4472	4939	5552	6176	6767	7369
	1.5	1.0	4668	5276	5893	6481	7085	7689
		0.8	4195	4775	5378	5986	6561	7153
	Mean	1.0	4790	5360	5925	6485	7065	7630
		0.8	4470	4995	5570	6150	6713	7285

* The calculation was made with $E = 356$ MPa.

Table A2.5. Calculated mean resistances to contact shear along the skin surface of injection piles (Type 6) for control-mix sands.

Soil	Pile diameter d_2, m	Index of soil density I_D	Value of $\bar{\tau}_s$*, kPa with pile length L, m					
			6	8	10	12	14	16
Coarse sand	0.15	1.0	297	301	306	311	315	319
		0.8	224	228	232	237	241	245
	0.20	1.0	271	276	281	285	289	293
		0.8	206	210	215	219	223	227
	0.25	1.0	252	257	262	266	270	275
		0.8	192	196	201	205	209	214
Medium sand	0.15	1.0	176	180	185	190	194	198
		0.8	108	112	117	121	125	130
	0.20	1.0	160	164	168	172	177	182
		0.8	99	103	108	112	116	121
	0.25	1.0	150	154	158	162	167	171
		0.8	93	97	102	107	110	114
Fine sand	0.15	1.0	142	146	150	154	159	163
		0.8	105	109	113	117	122	126
	0.20	1.0	129	134	138	142	147	151
		0.8	95	100	105	108	113	117
	0.25	1.0	122	126	130	135	139	143
		0.8	90	94	88	103	107	111

* The calculation was made with $E = 356$ MPa.

Table A2.6. Calculated resistances under the lower ends of driven piles R_f for control-mix sands.

Soil	Pile section cm	Index of density of contact zone I_D	Value of R_f, kPa with pile length L, m					
			6	8	10	12	14	16
Coarse sand	25 × 25	1.0	10620	11100	11599	12008	12466	12936
		0.8	7963	8424	8864	9325	9735	10216
		0.6	5922	6393	6853	7325	7796	8256
	30 × 30	1.0	10611	11077	11113	11552	12497	12921
		0.8	7771	8254	8714	9196	9667	10139
		0.6	5798	6259	6741	7200	7661	8132
	35 × 35	1.0	10283	10779	11281	11816	12315	12826
		0.8	7493	7964	8436	8907	9389	9849
		0.6	5672	6161	6649	7148	7638	8126
	40 × 40	1.0	9978	10636	11057	11590	12148	12669
		0.8	7469	7958	8389	8819	9364	9845
		0.6	5532	6042	6539	7060	7569	8066
	Mean	0.8	7575	8150	8600	9060	9540	10010
		0.6	5730	6215	6695	7180	7665	8145
Medium sand	25 × 25	1.0	6090	6506	6913	7310	7858	8113
		0.8	4046	4452	4849	5255	5672	6069
		0.6	3148	3571	3995	4408	4841	5255
	30 × 30	1.0	5792	6209	6625	7031	7446	7844
		0.8	3908	4315	4711	5128	5534	5931
		0.6	3035	3468	3892	4315	4728	5162
	35 × 35	1.0	5756	6179	6602	7023	7416	7853
		0.8	3774	4190	4597	5122	5372	5807
		0.6	2942	3365	3748	4212	4625	5048
	40 × 40	1.0	5583	6017	6440	6863	7296	7710
		0.8	3724	4141	4558	4984	5369	5806
		0.6	2906	3333	3739	4186	4603	5040
	Mean	0.8	3863	4275	4680	5120	5485	5905
		0.6	3010	3435	3845	4280	4700	5125
Fine sand	25 × 25	1.0	5200	5598	6014	6391	6808	7205
		0.8	3908	4314	4711	4895	5500	5902
		0.6	2808	3198	3578	3959	4358	4646
	30 × 30	1.0	4936	5333	5749	6146	6553	6950
		0.8	3888	4295	4622	4982	5395	5788
		0.6	2715	3114	3494	3884	4265	14654
	35 × 35	1.0	4780	5176	5583	5990	6396	6803
		0.8	3698	4105	4521	4918	5324	5722
		0.6	2708	3079	3479	3888	4278	4678
	40 × 40	1.0	4727	5131	5554	5957	6371	6774
		0.8	3606	3994	4400	4797	5213	5620
		0.6	2661	3068	3465	3881	4287	4694
	Mean	0.8	3775	4175	4565	4900	5360	5760
		0.6	2725	3115	3504	3903	4295	4670

230 *Appendix 2*

Table A2.7. Calculated bearing capacity at lower end of driven piles F_f for control-mix sands.

Soil	Pile section, cm	Values of index of density I_D	Value of F_f kN with pile length L, m					
			6	8	10	12	14	16
Coarse sand	25 × 25	1.0	664	694	722	751	779	809
		0.8	498	527	554	583	608	639
		0.6	370	400	428	458	487	516
	30 × 30	1.0	955	997	1000	1040	1125	1163
		0.8	699	743	784	828	870	912
		0.6	522	563	607	648	690	732
	35 × 35	1.0	1260	1320	1382	1447	1508	1571
		0.8	918	976	1033	1091	1150	1206
		0.6	695	755	814	874	936	995
	40 × 40	1.0	1596	1702	1769	1854	1944	2027
		0.8	1195	1273	1342	1411	1498	1575
		0.6	885	967	1046	1130	1211	1291
Medium sand	25 × 25	1.0	380	407	432	457	491	507
		0.8	253	278	303	328	354	379
		0.6	197	223	250	276	303	328
	30 × 30	1.0	521	559	596	633	670	706
		0.8	352	388	424	462	498	534
		0.6	273	312	350	388	426	465
	35 × 35	1.0	705	757	809	860	908	962
		0.8	462	513	563	627	658	711
		0.6	360	412	459	516	566	618
	40 × 40	1.0	893	963	1030	1098	1167	1234
		0.8	596	662	729	797	859	929
		0.6	465	533	598	670	736	806
Fine sand	25 × 25	1.0	325	350	375	399	425	450
		0.8	244	270	294	306	344	369
		0.6	176	200	224	247	272	290
	30 × 30	1.0	444	480	517	553	590	625
		0.8	350	386	416	448	486	521
		0.6	244	280	314	350	384	419
	35 × 35	1.0	586	634	684	734	748	833
		0.8	453	503	554	602	652	701
		0.6	332	377	426	476	524	573
	40 × 40	1.0	756	820	888	953	1019	1084
		0.8	577	639	704	768	834	899
		0.6	426	491	554	621	686	751

Table A2.8. Calculated mean resistances to contact shear $\bar{\tau}_s$ along the skin surface of a trench foundation, and additional pressures at the level of the lower end for control-mix sands.

Soil	Length of foundation L, m	Density of soil	Value of $\bar{\tau}_s$*, kPa, with pile length L, m										
			6		8		10		12		14		16	
			$\bar{\tau}_s$	σ_{ad}	$\bar{\tau}_s$	σ_{ad}	$\bar{\tau}_s$	σ_{ad}	$\bar{\tau}_s$	σ_{ad}	$\bar{\tau}_s$	σ_{ad}	$\bar{\tau}_s$	σ_{ad}
Coarse sand	2.0	Dense	87	110	91	115	96	121	101	127	106	133	111	140
		Medium	64	81	69	87	74	93	78	98	83	105	88	111
	3.0	Dense	82	103	87	110	91	115	96	121	101	127	105	132
		Medium	61	77	65	82	69	87	73	92	78	98	83	105
	6.0	Dense	76	96	80	101	84	106	89	112	94	118	99	125
		Medium	57	72	61	77	65	82	69	87	74	93	79	100
Medium-grained sand	2.0	Dense	43	54	47	69	52	66	57	72	62	78	66	83
		Medium	33	42	37	47	42	53	46	58	53	67	58	73
	3.0	Dense	40	50	44	55	49	62	53	67	58	73	63	79
		Medium	31	39	35	44	40	50	44	55	49	62	56	71
	6.0	Dense	39	49	44	55	47	59	52	66	56	71	60	76
		Medium	30	38	34	43	38	48	43	54	47	59	52	66
Fine sand	2.0	Dense	42	53	45	57	50	63	54	68	59	74	63	79
		Medium	31	39	35	44	39	49	44	55	48	61	53	67
	3.0	Dense	38	48	43	54	47	59	51	64	56	71	60	76
		Medium	30	38	33	42	37	47	41	52	46	58	51	64
	6.0	Dense	38	48	42	53	45	57	50	59	55	69	58	73
		Medium	29	37	33	42	37	47	42	53	45	57	50	63

* The calculation was made with $E = 356$ MPa.

Table A2.9. Calculated mean resistances to contact shear $\bar{\tau}_s$ along the skin surface of a trench foundation with injected skin surface and additional pressures at the level of the lower end σ_{ad} for control-mix sands.

Soil	Length of foundation L, m	Index of soil density	Values of $\bar{\tau}_s$* and σ_s, kPa, with the height of the foundation H, m											
			4 noninject.	6		8		10		12		14		16	
			$\bar{\tau}_s$	$\bar{\tau}_s$	σ_{ad}	$\bar{\tau}_s$	σ_{ad}	$\bar{\tau}_s$	σ_{ad}	$\bar{\tau}_s$	σ_{ad}	$\bar{\tau}_s$	σ_{ad}	$\bar{\tau}_s$	σ_{ad}
Coarse sand	2.0	1.0	15	116	146	120	151	124	156	130	164	135	170	141	178
		0.8	14	99	125	103	130	107	135	113	142	117	147	122	154
	3.0	1.0	15	105	132	110	134	115	145	120	151	126	154	131	165
		0.8	14	94	118	98	123	102	129	107	135	112	141	116	146
	6.0	1.0	15	106	134	111	140	112	141	117	147	123	155	128	161
		0.8	14	88	110	91	115	95	110	100	126	105	132	110	139
Medium sand	2.0	1.0	12	74	93	78	98	82	103	86	108	91	115	95	120
		0.8	11	53	67	59	74	61	77	65	82	70	98	74	93
	3.0	1.0	12	69	87	73	92	77	97	81	102	86	108	90	113
		0.8	11	53	67	59	74	61	77	65	82	70	88	74	93
	6.0	1.0	12	67	84	72	91	76	96	80	101	84	106	88	111
		0.8	11	48	60	53	67	56	71	60	76	63	79	67	84
Fine sand	2.0	1.0	11	62	78	66	83	69	87	74	93	78	98	82	103
		0.8	11	51	64	55	69	59	74	63	79	67	84	72	91
	3.0	1.0	11	57	72	61	77	66	83	69	87	74	93	78	98
		0.8	11	47	59	51	64	55	69	60	76	64	81	68	86
	6.0	1.0	11	56	71	60	76	65	82	68	86	72	91	76	96
		0.8	11	47	59	50	63	54	68	58	73	62	78	67	84

* The calculation was made with $E = 356$ MPa.

Table A2.10. Calculated resistances under the lower end of a footing for control-mix sands.

Soil	Length of foundation L, m	Breadth of foundation b, m	Density of soil		Values of R_f^*, kPa, with the height of the foundation H, m					
					6	8	10	12	14	16
Coarse sand	2.0	0.6	Dense		2770	3131	3500	3869	4238	4616
			Medium		2153	2506	2859	3203	3565	3917
		0.8	Dense		2890	3276	3670	4066	4461	4865
			Medium		2347	2742	3136	3522	3926	4321
		1.2	Dense		2900	3294	3697	4100	4503	4916
			Medium		2366	2769	3172	3566	3979	4382
	3.0	0.6	Dense		2730	3117	3486	3864	4242	4611
			Medium		2129	2479	2828	3177	3535	3902
		0.8	Dense		2832	3244	3638	4041	4444	4838
			Medium		2126	2482	2837	3192	3555	3928
		1.2	Dense		2731	3137	3524	3927	4317	4709
			Medium		2127	2491	2854	3217	3593	3970
	6.0	0.6	Dense		2663	3036	3413	3800	4186	4590
			Medium		2075	2430	2779	3140	3504	3876
		0.8	Dense		2665	3052	3439	3835	4232	4638
			Medium		2083	2446	2810	3173	3545	3926
		1.2	Dense		2656	3054	3452	3859	4274	4692
			Medium		2069	2438	2807	3176	3554	3941
		Mean	Dense		2760	3150	3535	3930	4320	4720
			Medium		2165	2530	2900	3260	3640	4020
Medium-grained sand	2.0	0.6	Dense		1466	1751	2049	2341	2632	2917
			Medium		1265	1549	1841	2125	2437	2729
		0.8	Dense		1487	1740	2040	2332	2623	2909
			Medium		1266	1555	1850	2139	2455	2751
		1.2	Dense		1463	1756	2064	2365	2665	2959
			Medium		1228	1511	1801	2084	2395	2684
	3.0	0.6	Dense		1441	1730	2032	2331	2617	2912
			Medium		1252	1541	1837	2125	2428	2745
		0.8	Dense		1435	1727	2033	2325	2624	2923
			Medium		1224	1509	1798	2082	2380	2693
		1.2	Dense		1430	1722	2031	2326	2629	2939
			Medium		1199	1484	1768	2065	2344	2648
	6.0	0.6	Dense		1367	1651	1922	2214	2492	2770
			Medium		1223	1512	1800	2096	2385	2687
		0.8	Dense		1425	1734	2019	2330	2627	2924
			Medium		1209	1492	1778	2071	2358	2658
		1.2	Dense		1421	1729	2023	2338	2639	2940
			Medium		1200	1489	1778	2073	2362	2664
		Mean	Dense		1435	1725	2025	2320	2615	2910
			Medium		1230	1515	1805	2095	2395	2695
Fine sand	2.0	0.6	Dense		1407	1676	1960	2236	2519	2796
			Medium		1134	1396	1659	1928	2197	2465
		0.8	Dense		1386	1654	1936	2210	2491	2766
			Medium		1175	1450	1725	2006	2287	2569
		1.2	Dense		1372	1641	1924	2200	2483	2760
			Medium		1167	1443	1720	2002	2286	2569
	3.0	0.6	Dense		1360	1641	1916	2190	2479	2753
			Medium		1175	1450	1724	2006	2287	2568
		0.8	Dense		1349	1516	1933	2192	2485	2764
			Medium		1171	1443	1721	2000	2286	2571

Table A2.10. Continued.

Soil	Length of foundation L, m	Breadth of foundation b, m	Density of soil	Values of R_f*, kPa, with the height of the foundation H, m					
				6	8	10	12	14	16
		1.2	Dense	1335	1614	1886	2178	2471	2750
			Medium	1163	1435	1714	1993	2278	2564
	6.0	0.6	Dense	1338	1614	1884	2140	2450	2720
			Medium	1137	1409	1682	1960	2226	2505
		0.8	Dense	1265	1621	1895	2155	2469	2742
			Medium	1169	1442	1713	2010	2284	2571
		1.2	Dense	1319	1599	1872	2132	2447	2720
			Medium	1172	1429	1746	1997	2270	2558
		Mean	Dense	1350	1620	1910	2180	2475	2750
			Medium	1160	1435	1710	1990	2265	2550

*The calculation was made with $E = 356$ MPa.

Table A2.11. Calculated resistances under the lower injected end of a footing R_f for control-mix sands.

Soil	Length of foundation L, m	Breadth of foundation b, m	Density of soil I_D	Values of R_f*, kPa, with the height of the foundation H, m					
				6	8	10	12	14	16
Coarse sand		0.6	1.0	3453	3823	4192	4589	4968	5365
			0.8	2901	3260	3624	2998	4366	4736
	2.0	0.8	1.0	3477	3858	4238	4646	5036	5444
			0.8	3032	3418	3803	4207	4593	4997
		1.2	1.0	3516	3910	4304	4726	5129	5552
			0.8	2926	3208	3590	3981	4373	4774
		0.6	1.0	3326	2722	4108	4498	4899	5286
			0.8	2866	3235	3613	3991	4369	4738
	3.0	0.8	1.0	3352	3761	4160	4559	4977	5376
			0.8	2871	3248	3635	4028	4408	4785
		1.2	1.0	3350	3771	4183	4595	5026	5437
			0.8	2719	3080	3452	3824	4196	4559
		0.6	0.1	3373	3773	4126	4527	4947	5347
			0.8	2788	3165	3542	3928	4322	4710
	6.0	0.8	1.0	3366	3776	4137	4547	4976	5386
			0.8	2791	3183	3570	3966	4363	4769
		1.2	1.0	3369	3770	4140	4560	5000	5420
			0.8	2797	3190	3588	3995	4403	4820
		Mean	1.0	3400	3795	4175	4585	4995	5400
			0.8	2845	3220	3600	3990	4375	4765
Medium-grained sand		0.6	1.0	2015	2306	2597	2888	3194	3485
			0.8	1567	1855	2091	2410	2702	3000
	2.0	0.8	1.0	1982	2279	2575	2872	3176	3473
			0.8	1506	1794	2083	2386	2674	2963
		1.2	1.0	1952	2272	2577	2882	3188	3493
			0.8	1505	1813	2099	2393	2672	2966
		0.6	1.0	1978	2274	2570	2866	3169	3465
			0.8	1527	1822	2111	2400	2688	2984
	3.0	0.8	1.0	1977	2272	2582	2885	3195	3497
			0.8	1542	1841	2133	2425	2717	3016

Table A2.11. Continued.

Soil	Length of foundation L, m	Breadth of foundation b, m	Density of soil I_D	\multicolumn{6}{c}{Values of R_f*, kPa, with the height of the foundation H, m}					
				6	8	10	12	14	16
Fine sand	6.0	1.2	1.0	1980	2290	2600	2910	3228	3538
			0.8	1515	1819	2115	2411	2707	3010
		0.6	1.0	1818	2109	2387	2665	2943	3221
			0.8	1506	1812	2097	2389	2667	2959
		0.8	1.0	1823	2294	2606	2917	3229	3540
			0.8	1417	1816	2105	2402	2684	2981
		1.2	1.0	1948	2280	2595	2911	3191	3542
			0.8	1501	1817	2110	2411	2698	2998
		Mean	1.0	1940	2265	2565	2865	3565	3475
			0.8	1510	1820	2105	2405	2690	2985
	2.0	0.6	1.0	1700	1975	2243	2524	2799	3074
			0.8	1475	1749	2024	2299	2573	2861
		0.8	1.0	1638	1913	2194	2462	2743	3018
			0.8	1426	1701	1976	2264	2539	2813
		1.2	1.0	1601	1876	2157	2425	2707	2982
			0.8	1424	1702	1981	2274	2552	2831
	3.0	0.6	1.0	1618	1889	2166	2430	2707	2978
			0.8	1415	1686	1956	2240	2511	2781
		0.8	1.0	1614	1889	2171	2440	2722	2998
			0.8	1439	1721	2003	2299	2581	2863
		1.2	1.0	1609	1890	2177	2451	2738	3018
			0.8	1416	1696	1977	2271	2551	2832
	6.0	0.6	1.0	1607	1992	2165	2433	2709	2984
			0.8	1408	1677	1953	2228	2504	2786
		0.8	1.0	1606	1882	2164	2433	2708	2984
			0.8	1442	1718	2001	2284	2567	2857
		1.2	1.0	1626	1881	2172	2452	2737	3023
			0.8	1411	1690	1980	2265	2550	2843
		Mean	1.0	1625	1900	2180	2450	2730	3000
			0.8	1430	1704	1320	2270	2550	2830

* The calculation was made with $E = 356$ MPa.

Table A2.12. Calculated values of contact resistance to shear τ_s, kPa for flat reinforcement in soil depending on the ratio E/E_r.

Soil	Index of soil density I_D	Breadth of element b_r, m	Without allowing for dilatancy	Values E/E_r									
				0	0.01	0.05	0.10	0.50	1.00	2.00	3.00	5.00	8.00	10.00
Coarse sand	1.0	0.1	97	272	271	271	270	263	253	234	216	178	122	84
		0.2	97	240	239	239	237	231	221	203	184	146	90	53
		0.5	97	207	207	206	205	198	189	170	152	116	61	24
	0.6	0.1	87	169	169	168	167	160	150	131	113	75	19	–
		0.2	87	156	156	155	154	147	137	119	99	62	6	–
		0.5	87	141	141	140	139	132	122	104	85	48	–	–
Medium-grained sand	1.0	0.1	78	173	173	172	171	163	153	135	115	77	20	–
		0.2	78	156	156	155	154	147	137	118	99	62	5	–
		0.5	78	137	137	136	135	128	118	99	80	43	–	–
	0.6	0.1	71	103	103	102	106	93	84	65	45	7	–	–
		0.2	71	96	96	95	94	86	77	58	38	–	–	–
		0.5	71	91	91	90	89	82	72	53	34	–	–	–
Fine sand	1.0	0.1	70	146	146	145	144	136	127	108	88	50	–	–
		0.2	70	132	132	131	130	122	113	94	74	36	–	–
		0.5	70	118	118	117	116	109	100	83	65	30	–	–
	0.6	0.1	68	98	98	97	96	88	79	60	41	2	–	–
		0.2	68	92	92	81	90	82	73	54	34	–	–	–
		0.5	68	89	89	88	87	80	71	54	1	–	–	–

* The calculation was made with $\sigma_{n_o} = 100$ kPa; $E = 356$ MPa; $l/b = 50$ for control-mix sands.

References

Amsheus, I. & V. Culeshus 1982. Osobennosti opredeleniya soprotivleniya peschanykh gruntov sdvigu apparatam SPF-1. (Pecularities of definition of sandy soils shear resistance on apparatus SPF-2). *Conf. Geotehnika-5 Baltiiskikh respublik i Byelorussii (Conf. Geotechnics-5 of Baltic Republ. and Byelorussia). Byelorussian Polytechnical Institute, Minsk, pp. 83-86.*

Arz, P. & K. Krubasik 1986. Mantel und Fussverpressung bei Bohrpfahlen, *Phahlsymposium'86, Darmstadt.*

Baholdin, B.V. 1986. Experimentalnye i teoreticheskie issledovaniya protsessa vzaimodeistviya grunta s zabivnymi svayami i sozdanie na ikh osnove prakticheskikh metodov rastcheta svay (Experimental and theoretical investigation of the process of soil-driven piles interaction and creation on its base of practical methods to pile design). Dissertatsiya na stepen d.t.n. (D. Sc. Thesis), VNIIOSP, Moskow.

Barkan, D.D., Yu.G. Trofimenkov & T.N. Golubtsova 1974. O zavisimosti mezhdu uprugimi i prochnostnymi kharakteristikami gruntov (On dependence between elastic and strength characteristics of soils). Osnovaniya, fundamenty i mekhanika gruntov, 1 *(Journal of Bases, Foundations and Soil Mechanics, Moskow, 1)*, pp. 29-31.

Bauer Spezialitiefbau GmbH. Messungen im Spezialitiefbau. Bericht uber Phahlbelastungsversuche auf der Baustelle Mainz, Gymnasium am Kurfürstlichen Schloss. Materials granted by Prof. E. Franke, Baugrundinstitute Darmstadt.

Baugrundinstitutes Darmstadt. Bauvorhaben Gymnasium am Schloss, Mainz. Sondervorschlag-Fussverpresste Phahle. Materials granted by Prof. E. Franke Baugrundinstitute Darmstadt.

Bezimana O. 1989. Nesustchaya sposobnost armirovannogo grunta (Bearing capacity of reinforced earth). Dissertatsiya na stepen k.t.n. (Ph. D. Thesis). Byelorussian Polytechnical Institute, Minsk.

Bezukhov, N.I. 1968. Osnovy teorii uprugosti, plastichnosti i polzuchesti (Bases of theory of elasticity and creeping). Visshaja Shcola publ., Moskow, pp. 428-432.

Bishop, A.W. 1972. Shear strength parameters for undisturbed and remoulded soil specimens. Stress-strain behaviour of soils (R.N.G. Parry (ed)). *Proc. of the Roscoe Memorial Symposium, Cambridge University*, pp. 3-58, 134-139.

Bishop, A.W. & Henkel D.J. 1962. The measurement of soil properties in the triaxial test. Edition Arnold.

Bromvell, L.G. 1966. The friction of quartz in high vacuum. M.I.T. Dept. of Civil Engineering, Research Report B66-18.

Burgov, A.K., R.M. Narbut & V.P. Sipidin 1987. Issledovanie gruntov v usloviyah trekhosnogo szhatiya (Investigation on soils in conditions of triaxial compression. Stroyizdat publ., Moskow.

Cambefort, H. 1964. Injection des sols, tome 1. Principes et methods, Eyrolles, Paris.

Casagrande, A. 1935. New fact in soil mechanics from the research laboratories. Eng. News Record, Vol. 115, v. 10.

Casagrande, A. 1936. Characteristics of cohesionless soils affecting the stability of slopes and earth fills. Contribution to Soil Mechanics, Boston Soc., Civ. Engineers, pp. 257-276.

Černak, B. 1973. Effect of suspension on skin friction in sands: New approaches to problems of bearing capacity and settlement of piles. *Proc. 8th Int. Conf. on Soil Mechanics and Foundation Engineering, Moscow*, pp. 67-69.

Christian, C., J.-D. Rakotondramanitra & J. Bakot 1986. Soil reinforcement study making use of waste plastic materials with large triaxial cell. *Proc. XI Conf. Bases of Foundation Brno 1986, CSSR.*

Dalmatov, B. I., F. K. Lapshin & Yu. V. Rossikhin 1975. Proektirovanie svaynikh fundamentov v usloviyakh slabykh gruntov (Design of pile foundations on weak soils). Stroyizdat publ., Moskow.

Davidson, R. 1988. Novye ekonomichnye metody ustroystva fundamentov (New economical methods of foundation). Grazhdanskoe stroitelstvo, No. 4. (Journal of Civil Engineering No. 4), Moskow, pp. 4-9.

DDR-Vorschrit. Unterirdische Wände, Heft 51 der Schriftenreihe der Bauforschung, Reine Ing. -und Tiefbau Vornorm, Teil II.

Dranovski, A. N. 1985. Parametry prochnosti razuprochnyayustchikhsya pri sdvige peschanykh gruntov (Strength parameters of sandy soils loosing strength while shear). V sbornike 'Issledovanie raboty osnovaniy i Fundamentov v slozhnykh gruntovykh usloviyakh' (In book 'Investigation of work of bases of foundations in complicated soil conditions'). J. Kazan Polytechnical Institute, Kazan, pp. 50-55.

Dranovski, A. N. & S. N. Rossikhin 1983. Issledovanie rasporynkh napryazheniy pri stesnionnom sdvige gruntov (Investigation of thrust stresses while constrained shear of soils). *Conf. 'Osnovaniya i fundamenty v slozhnykh inzhenerno-geologicheskikh usloviyakh' (Proc., Conf. 'Bases and foundations in complicated engineering-geological conditions') Kazan Polytechnical Institute, Kazan*, pp. 17-21.

Dranovski, A. N. & M. S. Vorobiov 1968. Issledovanie rasporynkh napryazheniy pri stesnionnom sdvige gruntov (Investigation of soil strength under constrained shear). Conf. 'Geotekhnika Povolzhya-3' (Conf. Geotechnics of Near-Volga Region-3). Kazan Polytechnical Institute, Kazan, pp. 35-39.

Drescher, A. & G. de Josselin de Jong 1972. Photoelastic verification of a mechanical model for the flow of a granular material. *Journal of the Mechanics and Physics of Solids*, 2, pp. 337-351.

Ellner, A. 1981. Einfluss des Herstellungsverfahrens von Grossbohrpfählen auf die Tragfähigkeit Ver offentlichungen des Grundbauinstitutes der Landes-Gewerbeanstalt Bayern, Helt 50, Eigenverlag LGA, pp. 51-63.

El-Sohby, A. 1969. Deformation of sands under constant stress ratio *Proc. 7th Int. Conf. Soil Mechanics and Foundation Engineering, Mexico*, Vol. 1, pp. 111-119.

Erikhov, B. P. 1961. Dinamicheskiy metod opredeleniya modulya sdviga myagkikh gruntov estestvennoy structury (Dynamical method for estimation of modulus of elasticity of soft soils of natural structure). Trudy VNIIG im. Vedeneeva (*Proc. of VNIIG Institute*), Leningrad, pp. 17-25.

Farmer, I.W., P.I.C. Buckley & Sliwinski 1971. The effect of bentonite on the skin friction of cast-in-place piles. Behaviour of piles, Institute of Civil Engineering, London, pp. 115-120.

Fernandez-Renau, L. 1965. Discussion in Session 6, Deep Foundations. *Proc. 6th Int. Conf. Soil Mechanics Montreal*, pp. 495-296.

Filonenko-Borodich, M.M. 1961. Mekhanicheskie teorii prochnosti (Mechanical theories of strength). Izdatelstvo Moskovskogo Universiteta (Publ. of Moskow University), Moskow.

Geffen, S.A. & M. Amir 1971. Effect of construction procedure on load-carrying behaviour of single piles and piers. *Proc. 4th Asian Regional Conf. SMFE, Bangkok*, Vol. 1, pp. 263-268.

Gersevanov, N.M. & D.E. Polshin 1948. Osnovy mekhaniki gruntov i ikh prakticheskie prilozheniya (Bases of soil mechanics and their practical uses). Gosstroyizdat publ., Moskow, pp. 124-129.

Goldstein, M.N. 1979. Mekhanicheskie svoistva gruntov (Mechanical properties of soils). Stroyizdat publ., Moskow.

GOST 5180-75. Soils. Method of laboritory estimation of humidity.

GOST 5182-78. Soils. Method of laboratory estimation of unit weight.

GOST 10180-78. Concretes. Methods of estimation of strength to tension and compression.

GOST 12248-78. Soils. Methods of laboratory estimation of shear resistance.

GOST 12536-79. Soils. Methods of laboratory estimation of grain composition.

GOST 25100-82. Soils. Classification.

Grutman, M.S. 1969. Svaynie fundamenty (Pile foundations). Budivelnik publ., Kiev, p. 42.

Hanna, T.H. 1982. Foundation in tension. Ground Anchors, Trans Tech. Publ., McGraw-Hill Book Co.

Hardy, W.B. & J. Bircumshaw 1925. Boundary lubrication – Plane surfaces and the limitation of Amonton's law. *Proc. Roy. Soc.*, A 108, pp. 1-27.

Hobst, L. 1987. Gruntovye ankery vysokoi nesustchey sposobnosti (Soil anchors of high load-holding capacity). *Arkhitectura i stroitelstvo Belorussii (Journal of architecture and Construction of Byelorussia)*. Minsk, pp. 29-31.

Hobst, L. & J. Zajic 1983. Anchoring in rock and soil. Elsevier Publ. Co.

Hong Kong Specification 1980. Guide and specification for ground anchors.

Hope, V. 1989. Estimation of shear modulus by geophysical method. *Proc. 3 Young geotech. eng. conf., additional volume, Raubichi, USSR*.

Horn, H. & D. Deere 1962. Friction characteristics of minerals. *Geotechnique*, Vol. 12, No. 4, pp. 319-335.

Hvorslev, M.J. 1937. Uber die Festigkeitseigen Shaften gestorter Bindiger Boden, Ingvidensk, Skr. A, No. 45 (English trans. No. 69-5, Waterways Experiment Station, Vicksburg, Miss., 1969).

ISSMFE European Pile Committee. Suggestion, encl. 1: How shall safety requirements for vertically loaded single piles be defined.. Inst. fur Grundbau, Boden un Fels-mechanik, Darmstadt.

Ivanov, P.L. 1985. Grunty i osnovaniya gidrotekhnicheskikh sooruzheniy (Soils and bases of hydrotechnical structures). Visshaya shkola Publ., Moskow.

Ivering, J.W. 1981. Developments in the concept of compressed tube anchors. *Ground Engineering*, Vol. 14, No. 2.

Jaropolsky, I.V. 1933. K voprosu o koeffitsiente treniya v peskakh (On the question of coefficient of friction in sands). Trudy LIVT (*Proc. of Leningrad Institute for Water Transport*), Leningrad, pp. 18-37.

Kenney, T. 1967. Shear strength of soft clay. *Proc. Geotech. Conf.,* Vol. 2, pp. 49-55, Oslo.
Kempfert, H.G. 1982. Report about load tests on small-diameter pressure-grouted piles. Institute für Grundbau und Bodenmechanik, Technische Universität München.
Kienberger, H. 1975. Diaphragm walls as load-bearing foundations, diaphragm wall and anchorages. Institute of Civil Engineering, London, pp. 19-21.
Koreck, H.W. 1976. Small diameter bore injection piles. *Ground Engineering,* Vol. 9.
Koreck, H.W. Reports about load tests on small-diameter pressure-grouted piles. (Partly not publ.), Institut für Grundbau und Bodenmechanik, Technische Universität München.
Lambe, T.W. & R.V. Whitman 1977. Mechanika gruntow, tom 1, Warszawa, Arkady, pp. 164-165. (Soil Mechanics. John Wiley and Sons, 1969).
Lapshin, F.K. 1987. Rastchet osnovanii odinochnich svay na verticalnyu nagruzku (Design of bases of single piles on vertical load). Dissertatsiya na stepen d.t.n. (D.Sc. thesis). Saratov Polytechnical Institute, Saratov.
Malishev, M.V. 1980. Prochnost gruntov i ustoichivost osnovaniy sooruzheniy (Strength of soils and stability of bases for structures). Stroyizdat publ., Moskow.
Maslov, N.N. 1982. Osnovy mekhaniki gruntov i inzhenernoy geologii (Bases of soil mechanics and engineering geology). Stroyizdat publ., Moskow, pp. 465-477.
Mastrantuono, C. & A. Tomolio 1977. First application of a totally protected anchorage. *Proc. IX Int. Conf. on Soil Mech. and Found. Eng., Tokyo.*
Meissner, H. 1982. Tragverhakten und horizontal belasteter Bohrpfähle in Kornigen Boden (1). – Geotechnik, Deutsche Gesellschaft für Erd-und Grundbau, e. V., Essen, 5, pp. 1-13.
Mishakov, V.A. 1980. O protsesse uplotneniya peschanogo grunta pri ustroystve inektsionnykh ankerov (On the process of compaction of sandy soil during construction of injection anchors). *Zhurnal 'Spetsialnie stroitelnye raboty' No. 12 (Journal of Special Construction Works, No. 12)* Moskow, pp. 17-19.
Nadai, A. 1963. *Theory of flow and fracture of solids,* Vol. 2. McGraw-Hill Book Co. 1963.
Nikitenko, M.I. & D.Yu. Sobolevsky 1984. Inektiruemye tscmentnye rastvory (Injection cement grouts). *Journal of Metrostroy,* Moskow, No. 6.
Nikitenko, M.I. & D.Yu. Sobolevsky 1986. Unstability of soil characteristics at grouting. *Proc. 14 conf. 'Bases of structures – Brno 1986' CSSR,* pp. 334-338.
Nikolaevski, V.N. 1975. Posleslovie. Sovremennye problemy mekhaniki gruntov (Afterward. Modern problems of soil mechanics. Basic laws of soil mechanics). Seriya 2. Opredelyayustchie zakony mekhaniki gruntov (Serie 2`Basic laws of soil mechanics). Mir publ., Moskow, pp. 210-227.
Ostermayer, H. 1974. Construction, carrying behaviour and creep characteristics of ground anchors. *Proc. conf. on diaphragm walls and anchorages, Sept., 1974, Institute of Civil Engineers, London, 1975,* pp. 141-151.
Ostermayer, H. & F. Sheele 1977. Research on ground anchors in non-cohesive soils. *Proc. Session No. 4, Int. Conf. Soil Mech. and Found. Eng., Tokyo,* pp. 92-97.
Patie, D. 1963. Distribution of stresses in soil creation from action of shear forces along the skin surface of large diameter piles. *Proc. Int. conf. on soil mechanics and foundation engineering, Budapest.*
Popov, O.V. 1989. Nesustchaya sposobnost inektsionnych ankerov v dilatiruyustchikh nesviaznykh gruntakh (Load-holding capacity of injection anchors in dilating non-cohesive soils). Dissertatsiya na stepen k.t.n. (Ph.D. thesis), Byelorussian Polytechnical Institute, Minsk.
Reese, L.C., M.W. O'Neill & G.T. Touma 1983. Bored piles installed by slurry displacement. *Proc. 8th Int. Conf. Soil Mech., Moskow,* Vol. 2, pp. 1, 203-209.

Reference Book ... (1962). Spravochnik gidrogeologa (Reference Book on Hydrology), Moskow, pp. 130-138.
Reynolds, O. 1885. On the dilatancy of media composed of rigid particles in contact with experimental illustrations. *Philosophical Magazine and Journal of Science (Fifth Series)*, Vol.20, No. 127, pp. 469-481.
Reynolds, O. 1886. Experiments showing dilatancy, a property of granular material. *Proc. Roy. Inst.*, 2, pp. 354-363.
Rizhov, A.M. 1976. Opredelenie prochnosti i deformativnosti gruntov v stroitelstve (Definition of strength and deformativity of soils in construction). Budivelnik publ., Kiev.
Rizhov, A.M. & V.P.Vikharev 1959. Sluchay razzhizheniya peska v poimennoy nasypi (Case of sand liquefaction in flood-land fill). Sbornik 'Voprosy geotekhiki' (In book 'Questions of geotechnics'), Dnepropetrovsk, pp. 243-260.
Rowe, P.W. 1964. Notes on the relative principal strains of particle assembly during change in mean principal stress at constant ratio in the triaxial cell. Manchester University.
Rowe, P.W. 1972. Theoretical meaning and observed values of deformation parameters for soil. *Proc. of the Roscoe Memorial Symposium, Cambridge University, 1972*, pp. 143-192.
RSN 1987. Proektirovanie i ustroystvo transheinykh i svaynykh sten metodom 'stena v grunte'. Respublikanskie stroitelnye normy (Design and construction of diaphragm and pile walls. Republican construction norms). Gosstroy, Minsk.
Scheller, P. & M. Stocker 1983. Messungen bei statischen Phahlprobebelastungen Stand der Technik, sondergruch (seuszugsweise) aus dem Vortragsband zum Symposium Messtechnik in Erd- und Grundbau, München, 1983, Deutsche Gesellschaft für Erd- und Grundbau e. V. Essen.
Schnabel 1982. Tiebacks in foundation engineering and construction. New York: McGraw Hill, p. 157.
Sidorov, H.H. & V.P. Sipidin 1982. Sovremennye metody opredeleniya kharakteristik meknanicheskikh svoistv gruntov (Up-to-date methods for estimation of mechanical characteristics of soils). Stroyizdat publ., Leningrad, p. 136.
Skinner, A.E. 1969. A note on the influence of interparticle friction on the shearing strength of a random assembly of spherical particles. *Geochemique*, Vol. 19, No. 1, pp. 150-157.
Sobolevsky, D.Yu. 1984. Ushirenie skvazhin pri ustroystve buroinektsionnykh ankerov v peske (Widening of bore holes during construction of injection anchors in sand). Trudy LISI (*Proc. of Leningrad Engineering Construction Institute*), Leningrad.
Sobolevsky, D.Yu. 1985. Nesustchaya sposobnost buroinektsionnykh ankerov v nesviaznom grunte (Load-holding capacity of bore-injection anchors in non-cohesive soil). Dissertatsiya na stepen k.t.n. (Ph. D. thesis). Leningrad Engineering-Construction Institute, Leningrad.
Sobolevsky, D.Yu. 1986. Hydrodynamic illustration of injected bored anchor roots. *Proc. 14th conf. 'Bases of structures – BRNO 1986', CSSR*, pp. 339-343.
Sobolevsky, D.Yu. 1989. Design model of the frictional soil dilatancy. *Proc. 3 Young Geotech. Eng. Conf., Raubichi, USSR*.
Sobolevsky, D.Yu. & O.V. Popov 1987. Soprotivlenie sdvigu nesvyaznogo grunta po bokovoy poverkhnosti inektsionnykh ankerov i svai (Shear resistance of non-cohesive soils on the skin surface of injection anchors and piles). Sbornik 'Osnovaniya i fundamenty v geologicheskikh usloviyakh Urala' (*Proc. 'Bases and foundations in geological conditions of Ural'*), Perm Polytechnical Institute, Perm., pp. 78-85.
Sobolevsky, Yu. A. 1975. Vodonasystchennye otkosy i osnovaniya (Saturated slopes and

bases). Visheyshaya shkola publ., Minsk, pp. 371-373.

Soletanche 1990. Europe's largest aluminium works. *Construction Industry International*, Vol. 16, No. 10, pp. 14-16.

Soos, P. 1972. Anchors for carrying heavy tensile load into the soil. *Proc. 5th European Conf. Soil Mech., Madrid, 1972*, Vol. 2, Part 1, pp. 203-209.

Specifications ... 1975. Specifications for cast-in-place piles formed under bentonite suspension. 'Federation of Piling Specialists' Ground Engineering, March, Vol. 8.

Spivak, A.I. & A.I. Popov, 1975. Mekhanika gornykh porod (Mechanics of Rocks). Nedra publ., Moskow, p. 200.

Stelle, H. 1976. Behaviour of anchored sheet pile walls. Thesis Royal Inst. of Technology, Stockholm.

Stocker, M. 1977. Der Einfluss von Bentonitsuspension auf die Tragfähigkeit unverrohrt hergestellter Bohrpfähle. Sonderdruck (auszugsweise) aus dem Vortragsband zum Symposium Stand von normung, Bemessung und ausfukrung von Pfählen und Phählwänden, München 1977. Deutsche Gesellschaft fur Erd-und Grundbau e. V., Essen, 1979.

Stocker, M. 1980. Vergleich der Tragfähigkeit unterschiedlich hergestellter Pfahle. Vortagsband Baugrundtagung in Mainz, Deutsche Gesellschaft für Erd-und Grundbau e. V., Essen, pp 565-590.

Stocker, M. 1983. The influence of postgrouting on the load-bearing capacity of bored piles. *Proc. 8th European conf. on Soil Mechanics and Foundation Engineering, Helsinki, 1983*, pp. 167-170. Rotterdam: A.A. Balkema.

Stocker, M. 1986. Der Schneckenortbeton – Phäl, Kurz SOB – Pfähl. Phahlsymposium'86, Darmstadt.

Stocker, M. ..... Report about load tests on large-diameter bored piles and small-diameter pressure-grouted piles. Not publ., Karl Bauer Spezialtiefban, Schrobenhausen.

Stocker, M. & K. Bayer 1977. Vergleich der Tragfähigkeit unterschiedlich hergestellter Pfähle. Karl Bauer Spezialtiefbau, Schrobenhausen.

Stoetzer, E. Bored pile foundation. Construction Industry International, October, Vol. 13.

Šimek, J. & J. Verfel et al. 1989. The improvement of the pile bearing capacity. *Proc. XII International Conf. on Soil Mechanics and Foundation Engineering, Rio de Janeiro*, Vol. 2, pp. 1031-1034.

Terzaghi, K. & R. Peck 1967. Soil mechanics in engineering practice, 2nd edition. New York: John Wiley and Sons.

Tong, P.Y.L. 1970. Plane strain deformation of sands. Ph. D. thesis, University of Manchester.

Tschebotariof G.P. 1951. Soil mechanics, foundations and earth structures. McGraw Hill Book Co.

Tsitovich, N.A. 1963. Mekhanika gruntov (Soil mechanics). Gosstroyizdat publ., Moskow.

Wernick, E. 1977. Stresses and strains on the surface of anchors. *Proc. Session No. 4, 9th Int. Conf. Soil Mech. Found. Eng., Tokyo*.

Wernick, E. 1979. Tragfähigkeit zylindrischer Anker in Sand unter besonderer Berucksichtigung des Dilatanzverhaltens. Heft 75 des Institutes fur Bodenmechanik und Felsmechanik der Universität Karlsruhe, p. 148.

Subject index

Angle of
  contact friction 57
  internal friction 7, 78, 81, 91
  intergranular friction 7
  gear 7
  mineral friction 7, 8
Apparatus
  BCB-25, 41
  contact shear 34
  direct shear 39
  reinforced elements 42
Anchor
  construction 174
  design scheme 176
  load-holding capacity 182
  group effect 183
  root compression 179
  root strength 181
  root tension 177

Base
  resistance 156
Bulge
  external 15
  internal 15, 29, 86, 153

Coefficient of
  elastic resistance 24, 29, 150, 167
  form 24
  grain size & density 57, 59, 62, 81
  mineral friction 8

Poisson's 21, 24
  reliability 183
  root deformability 176, 179, 180
Cohesion of
  gear 24
Condition
  stress-deformative 138

Deformation
  dilatant 14, 70, 82, 93
  elastic 12
  plastic 12
Density 9
  critical 49, 61
  index 49, 61
Dilatancy
  constrained 17
  free 16
  process 14
  term 5
Dinamometer
  calibrating 38
  regulating 37

Failure
  area 106
  character 13
  mechanisms 6, 16, 24, 27, 29
Filtration
  contact 200
Friction
  contact 114, 118, 123, 128

intergranular 6
mineral 6
Footing
  design scheme 168, 170
  classification 112
  construction 115
  load-holding capacity 167

Grain
  gear 6
  packing 10
  strength 69

Injection 124, 132
  body 133

Length
  critical 180
Liquefaction 195
  model 197, 199

Massif
  deformation 98, 103
  elasticity 49, 56, 91, 98
Modulus of
  deformation 32
  elasticity 24, 28, 32
Moisture
  content 63
Model
  assumptions 23
  contact shear 24
  internal bulge 29
  phenomenological 24,

243

   27, 29
 'soil to soil' shear 27

Pile
 auger 119
 bore 113, 115
 classification 112
 compressed base 124, 158
 design scheme 151, 154
 injection 131, 161
 load-holding capacity 150, 160, 164
 with soil displacement 135, 162
Pressure
 concrete 141
 initial critical 54

Reinforcement
 continuous 188
 discrete 188
 shear resistance 191
 stone constructions 205

Reinforced earth
 bearing capacity 187
 concrete 206
 stress condition 189
 structures 187
Resistance to
 contact shear 51, 96
 direct shear 77, 96
 triaxial compression 101
Resistance
 active 101, 102
 base 130, 160, 164
 passive 101, 102
 skin (contact) surface 114, 152, 160, 176

Sand
 control-mix 50
Seismic
 activity 203
Shear
 contact 49
 direct 75
 mechanisms 6

 resistance 96
 surface 98
Strain
 dilatant 65
Strength
 dilitant component 51, 56, 78
 grain 69
Strength condition
 Coulomb 8, 51, 85
 Mohr 8, 88, 95
Stress
 dilatant 17
 parameter 20
 principal 90

Test method
 contact shear 38
 direct shear 41
 reinforced elements 43
 triaxial 46
Tunneling 202